HOW TO MODEL IT

PROBLEM SOLVING FOR THE COMPUTER AGE

HOW TO MODEL IT

PROBLEM SOLVING FOR THE COMPUTER AGE

Anthony M. Starfield
University of Minnesota

Karl A. Smith
University of Minnesota

Andrew L. Bleloch
Cambridge University

McGRAW-HILL PUBLISHING COMPANY

New York St. Louis San Francisco Auckland Bogotá Caracas
Hamburg Lisbon London Madrid Mexico Milan Montreal New Delhi
Oklahoma City Paris San Juan São Paulo Singapore Sydney Tokyo Toronto

This book was set in Times Roman by the College Composition Unit
in cooperation with Waldman Graphics, Inc.
The editors were B. J. Clark and Margery Luhrs;
the production supervisor was Janelle S. Travers.
The cover was designed by Carla Bauer.
R. R. Donnelley & Sons Company was printer and binder.

HOW TO MODEL IT
Problem Solving for the Computer Age

1 2 3 4 5 6 7 8 9 0 D O C D O C 8 9 4 3 2 1 0 9

ISBN 0-07-005897-0

Library of Congress Cataloging-in-Publication Data

Starfield, A. M.
How to model it: problem solving for the computer age/Anthony M. Starfield, Karl A. Smith, Andrew L. Bleloch.
p. cm.
Includes index.
ISBN 0-07-005897-0
1. Computer simulation. 2. Mathematical models. 3. Problem solving. I. Smith, Karl Aldrich. II. Bleloch, A. L. (Andrew L.) III. Title.
QA76.9.C65S74 1990
003—dc20 89-2716

ABOUT THE AUTHORS

Anthony M. Starfield is an applied mathematician. For over twenty years he has had two main interests: using computers to solve practical problems (from mining engineering to conservation biology) and teaching others to do the same. He currently divides his time between an appointment in Civil & Mineral Engineering and another in Ecology & Behavioral Biology, both at the University of Minnesota.

Karl A. Smith is an Associate Professor in the Department of Civil and Mineral Engineering at the University of Minnesota. His bachelor's and master's degrees are in metallurgical engineering and his Ph.D. is in educational psychology. He has published widely in both metallurgical engineering and engineering education; teaches courses on mineral processing, the application of operations research, and problem formulation and modeling; and conducts seminars on building small expert systems, cooperative learning, and structured controversy.

Andrew L. Bleloch is a physicist at the Cavendish Laboratory, Cambridge, England. He and Tony Starfield have collaborated on building and using models for the past eight years. Their preference for working in African game parks led to a previous book on modeling: *Building Models for Conservation and Wildlife Management*.

CONTENTS

PREFACE

*"For those, like me, who are not mathematicians, the computer can be a powerful friend to the imagination. Like mathematics, it doesn't only stretch the imagination. It also disciplines and controls it."**

—Richard Dawkins

Have you ever built a model?

Before you answer "No, that is why I am looking at this book," think again. You have surely been building conceptual models (in your mind) ever since you can remember—and prior to that too. You would not be able to think if you were incapable of building models. What you might not have done yet is build a model *explicitly* so that other people can understand it and perhaps use it.

Richard Dawkins conveys some of the excitement of what one can learn and accomplish by building explicit models. An explicit model is a laboratory for the imagination. You can tweak a model to see how it responds. You can argue whether the threads of logic really do knit together in a consistent fashion. You can explore its strengths and limitations. You can even guardedly make predictions and then argue how good (or poor) those predictions might be.

Do you think that it requires mathematical skills to do all these things? Is that perhaps why you have not tried to build models explicitly?

Once upon a time mathematics was indeed essential for building and exercising models, but the computer, as Richard Dawkins has discovered, has changed all that. In particular, personal computers with powerful but easily learned software (such as spreadsheets) have brought explicit modeling within everybody's reach; you don't even have to know much about computers.

This is the good news: everybody should be able to build computer models. The bad news is that a lot of people are building models and using them to make important decisions and predictions, without ever having learned to criticize a model or distinguish a good model from a bad one. We all know how easy

**The Blind Watchmaker*, by Richard Dawkins, London, Longman Scientific and Technical, 1986.

it is to misuse statistics; there are books and university courses to teach us how to apply statistics sensibly. In exactly the same way there is a growing need for books and courses that develop the art, skills, and discipline of modeling.

We have written this book to demonstrate that there indeed is a subject called "modeling" which can be learned and needs to be learned. We believe it is an important subject, not only because models are becoming so pervasive, but also because the skills of modeling are so closely related to the more general skills of problem solving. Learning to model is bound up with learning to solve problems and to think imaginatively and purposefully.

This is not a book about mathematics. Some may wish that there were more mathematics in it. However, our primary objective is to nurture good modeling practice. We hint, now and then, that mathematics can make for more elegant and powerful modeling, but our intent is to encourage those who have an interest in mathematics rather than to discourage those who do not. The skills we emphasize are not mathematical skills.

This book is also not about computing, although we expect the reader to be able to make calculations that would take a great deal of time without the help of a computer. We do not ask you to learn to write computer programs, but if you are not already working on a computer, we hope this book will provide the impetus for you to ask a friend to show you how to use standard computer application packages. You will find some useful references at the end of Chapter 1.

This *is* a book about learning how to model and how to develop methods for improving the way in which you learn to model. Notice how we deliberately use the verb "learn" rather than "teach." That is because we believe one learns about modeling by doing it, by reflecting on what one has done, and by discussing or comparing approaches and solutions with others. Our book is designed to promote this kind of learning.

It is not easy to create an active learning environment in something as passive as a book. We have nevertheless tried to adapt the approach we use in the classroom with first-year students in mathematics, engineering, and the sciences.

In the classroom we ask students to work in groups of two to four. We give them a problem. We don't tell them what is required to solve it. Instead we have them spend 20 or 30 minutes discussing its formulation (what questions to ask, what assumptions to make, what would be an acceptable type of answer, and so on) in their groups. We then randomly call on individuals to present their group's approach. Often we streamline the reporting by giving each group a transparency and pen so that they can describe their approach without having to transcribe it on the chalkboard. Two or three individuals are randomly called to report, and then an open invitation is made for other presentations that offer a different formulation.

These presentations inevitably provide material for discussion. We then work with this material, comparing, contrasting, and discussing what needs to be represented and what is at the heart of the problem. Students develop more interest because they have been struggling with the problem. It is *their* problem. They have made an investment and are eager to learn.

At this stage in the class we generally provide more information. We do this through lecture sometimes, but more often through questioning, discussion, or structured tutorial wherein we help the students provide the information.

The students then go away to work on the problem in their groups. Depending on the problem, they might report back (either verbally or in short written reports) on specific subtasks, or they may work through to their final written report. Once these have been submitted we have a final session in the classroom where, again, the students make short presentations and we compare solutions, point out the lessons we have learned, and perhaps talk about how we would have tackled the problem.

Each chapter in this book addresses a problem in a similar way: we give you modeling tasks and we reflect, with you, on what you might have done or how we would have tackled those tasks. You will probably learn something from this book by just reading it, but you will learn a great deal more if you conscientiously (and, we hope, enthusiastically) perform the tasks. We hope you will take the time to do this, because learning occurs when *you* do the thinking, make the decisions, and find the means to solve the problems. We want you to struggle with that uncomfortable feeling of taking risks in exploring the unknown. To be temporarily perplexed is a natural part of modeling.

You could probably get even more out of this book by finding a friend or group of friends to work with. Discuss the problems, explain yourselves, and take turns making presentations to one another. Preparing to teach is the surest way to learn.

It should be apparent by now that this is no ordinary textbook; first of all it is not aimed at a standard university course, and second it does not present material in the way in which material is usually "taught." Where then does the book fit?

We hope that a number of people (of all ages) will be able to (and will want to) read and work through the book just for the fun of it. We also believe that this could be a stimulating text for many university courses. However, our real hope is that it will spark the development of new courses that are unashamedly devoted to the subject of *modeling*.

Finally, there are a number of people we would like to acknowledge or thank. George Polya's book *How to Solve It* inspired our approach to problem solving and how to teach (sorry, *learn*) it. We have tried to acknowledge this in the title of our book. Doug Wilson, Alan Wassyng, and Sam Sharp all contributed to our interpretation of Polya's ideas and the development of our own.

Michael Sears and Alan Schoenfeld read the manuscript; we would like to thank them for their comments. Peter Samsel helped us prepare some of the diagrams. Peter Leach thought of the beer cooling problem in Chapter 4. The Board of Regents of the University of Minnesota gave one of us (AMS) the time (in the form of a single-quarter leave) to think about modeling, and Peter Jewell kindly helped to organize the place (Cambridge, England) to do it.

We would also like to thank the following reviewers for their comments and suggestions: Dermot Collins, University of Louisville; Arvid Eide, Iowa State

University; James Jacobs, Norfolk State University; and Gearold Johnson, Colorado State University.

B. J. Clark and Margery Luhrs transformed our manuscript into a book. We thank them and their colleagues at McGraw-Hill for their thoughtful and cheerful assistance. The figures were drawn using VP-Graphics™ with the help of Paperback Software.

Finally, we would like to thank all the students who participated, so enthusiastically, in our various modeling courses. We learned by watching you learn.

Anthony M. Starfield
Karl A. Smith
Andrew L. Bleloch

HOW TO MODEL IT

PROBLEM SOLVING FOR THE COMPUTER AGE

CHAPTER 1

INTRODUCING MODELS (AND THIS BOOK)

OUR OBJECTIVE

Modeling is a method used in disciplines as diverse as microbiology and macroeconomics. In fact modeling is an integral part of problem solving in *any* discipline. This book is concerned with helping you learn how to model.

Modeling is more like a craft than a science, and the process of learning should therefore be more like an apprenticeship than a course of study. Our objective in this book is to guide you through a modeling apprenticeship.

WHAT IS A MODEL?

A Descriptive Model

You almost certainly have a model in your mind that tells you what to expect from this book. It is probably based on your experience with books in general and textbooks in particular. It is quite possible that your model could be represented by something like Figure 1.1. Have a look at it and see if you agree. If not, take a few minutes to draw your own representation of this book and how you expect to use it.

Figure 1.1 represents a rather modest model in the sense that it does not try to accomplish very much. The model is based on experience rather than theory, and is descriptive rather than predictive. It does not, for example, tell you how much time you should set aside for studying the book. But it does have a purpose, namely, to anticipate what you expect to see and do with this book.

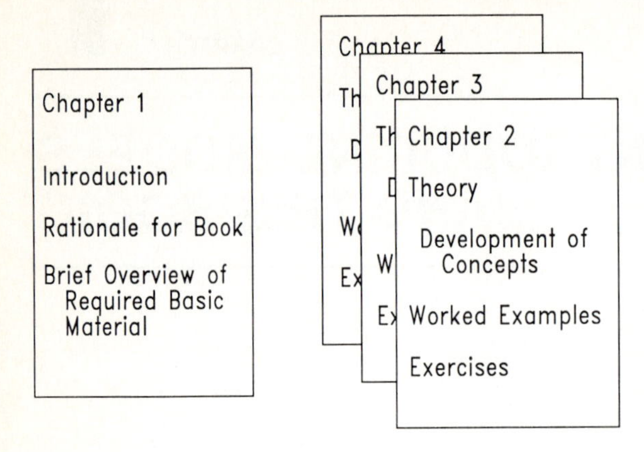

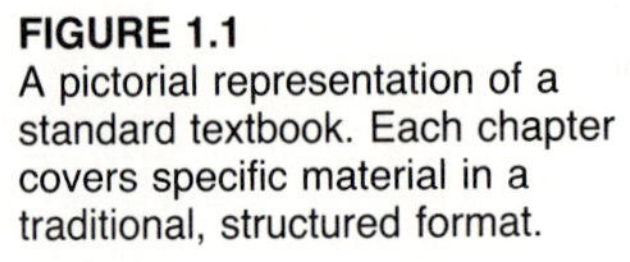

FIGURE 1.1
A pictorial representation of a standard textbook. Each chapter covers specific material in a traditional, structured format.

Is it a good model?

That is a difficult question to answer. What do we mean by the word "good"? Do we mean that the model is "true" in some absolute sense, or do we mean that it is "valid" in the sense that it has been checked successfully against a range of cases (perhaps by choosing books at random from the shelves in a college bookstore)? Before we become too embroiled in the details of defining a good model, it might help to anticipate a modeling heuristic (or rule of thumb) that we will be meeting in due course. That heuristic says: If a question is difficult to answer, start by asking a somewhat easier question.

In this case we can ask two somewhat easier questions:

1. Is the model in Figure 1.1 *useful* for anticipating what one would find in most texts?
2. How well does the model describe *this* particular text?

The answer to the first question is probably "yes." The answer to the second question is "not very well." As you will discover, we have not written this book in the standard textbook format.

Figure 1.2 is the model we used for planning the book. Look at it carefully and note the differences between it and Figure 1.1. Notice in particular that it is a model in which you, the reader, play a very active role.

A Predictive Model

Suppose we really do want to know how much time we will need to devote to this book. Whenever you see a ✔ displayed, such as the one below the next paragraph, you should recognize it as our way of telling you that we want you to *do* something before you read further. Please do not read ahead, because that would diminish your learning experience, but try to do what we ask of you in the time that we allocate. Sometimes you will see a ✓, which will refer you to a minor task that we ask you to stop briefly and perform.

In this case we will give you 10 minutes and ask you to estimate how much time *you* will need to work your way through this book.

✔

Did you read ahead expecting to find the answer? If you did, you are bound to be disappointed. *We* do not know how much time *you* will need!

However, we are very interested in how you went about making your estimate. Describe what you did in words, diagrams, equations, or whatever. Or better still, describe the model you used.

✔

Different people build different models, particularly when their circumstances are different. For all we know, you may have bought this book to while away the time on an air trip from New York to San Francisco, and so you may already have made up your mind to spend 5 hours with the book, and then throw it away. Your model, in this case, has nothing to do with the book or with what you hope to learn from the book, but it has a lot to do with the flying time from New York to San Francisco. From *your* perspective that is a perfectly good (in the sense of *useful*) model.

Alternatively, you may be one of those highly organized people who schedule their activities carefully in advance. Perhaps you have already scheduled

FIGURE 1.2
A flow diagram representing a chapter of this book. There are two essential activities: reading and performing tasks.

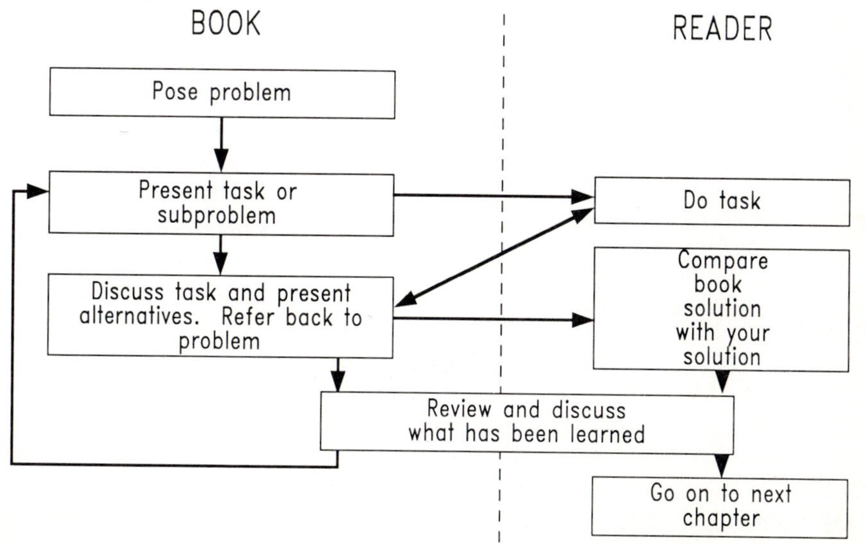

30 hours for this book. Again, your model in this case has nothing to do with the book, but a lot to do with all your other activities.

There are, however, some additional questions that you could ask yourself if you have approached time allocation in this way. The first is whether you really need 30 hours? If the answer to that is "yes," then you should be asking how much of the book you can absorb in the 30 hours. Both questions would lead you to a model that forces you to think about the book as well as yourself.

Most likely you began by looking at the book. Perhaps you looked at the number of pages and multiplied them by the time you usually take to read a page of print. You might even have introduced your own notation and written down an equation. For example, you might have said:

Denote the number of pages by P, and let p be the time to read a page. Let T represent the time to be devoted to the book. Then a simple model is

$$T = pP \tag{1.1}$$

In the space of 10 minutes you would have had time to look at the number of pages in the book (i.e., you would know P) and you could even have done a quick experiment to estimate p.

But is Equation 1.1 a good model? (Consider Figure 1.2 before answering.)

Figure 1.2 indicates that this book requires action on your part. Remember, each time you see the ✔ , we expect you to stop reading and *do* something. All those activities might take a lot more time than it takes to read the book.

One way to calculate that additional time would be to go through the book and estimate the time it will take to perform each of the tasks wherever you see a ✔ . This would improve the model, but it is not something that you could do in 10 minutes!

A quick way around that might be as follows: Suppose W is the average number of ✔ 's appearing on a page of print. If w is the average amount of time associated with a task, then Equation 1.1 could be expanded to

$$T = pP + wWP \tag{1.2}$$

In the time at your disposal you could have opened the book at random, counted pages until you had located, say, 10 ✔ 's, and kept note of the time allocated to each of them. From that information you could have estimated both w and W. Equation 1.2 would then be a reasonably good model, perhaps the best you could have built in the time allocated to you.

How does our model compare with your model?

That is not a rhetorical question. You might well have built a completely different model. It might not be as good (in the sense of useful and pertinent) as ours, but it could just as easily be better than our model. So, we really do want you to sit down and think about your model and our model and what you do like and do not like about each of them. But remember the guidelines: you had only 10 minutes in which to solve the problem, and we too thought of our solution to the problem in only a few minutes. When you evaluate models, you should always take into account the time (and other resources) allocated for their solution.

Bearing that in mind, write down a list of the points you like about our model, what you really like about your model, what you do not like about ours, and what, in retrospect, you do not like about yours.

✔

A CRITIQUE OF OUR MODEL

We have a slightly easier task than you; we do not know what your model looks like, so we can criticize only our own. But while you read the critique of our model, think about yours as well.

Do the same points apply? You should try to get into the habit of questioning and probing, because to do so is an essential component for building useful and reliable models.

What We Like about Our Model

We indicated earlier that every model must have a purpose. The purpose of this model is to estimate the time required to read this book.

A higher-level purpose from our point of view is to get you, the reader, engaged in the activity of modeling. We call this the metalevel purpose of the book.

We think we have done a good job of building a purposeful model. After all, Equation 1.2 is an equation to find precisely what we are looking for: the time T.

Given that we had only 10 minutes in which to build the model, we feel that we have done a competent job of isolating the most important issues. We asked ourselves the question: "How will the reader spend his or her time?" If our answer had only been "In reading the book," we would have missed the point of this particular book. By also answering "In doing the exercises," we have captured the two *essential* points. There are of course a whole lot more answers to the question, answers like "In looking carefully at the diagrams" or "In stopping to think about what we have written," but we did not have time to think about all these answers, and we feel that what we have ignored is not nearly as important as what we have included.

We like the fact that we introduced a notation and boiled our whole model down to a single equation. In the first place the notation makes it quite clear to the reader what we have put into our model and what we have left out of it. In the second place the equation is as explicit a recipe as one could wish to find for making the estimate. It tells the reader exactly how to calculate our estimate of the time T.

We nearly got hung up on the problem of how to estimate the time spent performing the various tasks. After all, each task has a different time allocated to it, and in 10 minutes it is just not possible to go right through the book and add up the times associated with each and every ✔ . It would have been simpler just to guess at the total amount of time the reader would spend on all the tasks, but that really would have been only a guess. By sampling a few pages, we have in effect built a much more reasoned (and reasonable) guess into our model.

Some Weak Points of Our Model

What are the weak points of the model? We cannot think of any really important weak points and that in itself is a very weak point. Most modelers tend to be complacent about their models and very often miss blatant weaknesses. If you cannot list the weak points of your model, then it is possible that the model is very good, but it is more likely that you have been unable to stand aside from your model to see its weak points.

So let us try a little harder to criticize our model.

Perhaps you brought up the point that we produced an equation, not an actual number for our answer. We did not say "The answer is 7 days 3 hours and 15 minutes," for instance.

We would argue that this is a strength rather than a weakness. We do not, for instance, have any idea of how fast you can read, and so we have introduced a symbol to represent your reading speed and left it to you to substitute the appropriate number. That is a more sensible approach than estimating the reading speed of an "average reader." We have built a model that takes into account (within the crude scope of the model) the special properties of the reader as well as those of the book.

"Now that you have mentioned the reader, do you think that you have paid sufficient attention to the reader as an individual? Perhaps the reader is a very speculative person, and you should have allowed for thinking time in your model." Is that what is in your mind?

If so, it is a good point, but we can accommodate thinking time without changing the model. For instance, we could argue that a thoughtful reader is just somebody who takes longer to read a page; all we need do is modify the reading speed p.

A more telling criticism is that we have not given any thought at all to the *accuracy* of our answer. There are no error bounds associated with Equation

1.2, and we have not considered how reliable our estimates of reading speed (p), the number of tasks per page (W), and the average time allocated to a task (w) might be.

There is not much we can do about this in the space of 10 minutes, but there are checks we could have suggested. For instance, estimate p by reading two pages. Comparing the times for each page will tell you something about a range of values (rather than just one value) for p.

How could you estimate a range of values for w and W? How would you improve your estimates of p, w, and W if you had 20 instead of 10 minutes?

Notice that if you had more time it might be more important to use the extra time for establishing error bounds rather than just improving the accuracy of the numbers themselves.

WHAT WE HAVE LEARNED SO FAR

Before we move on, let us pause to reflect on what we have done so far.

Creating an Active Learning Environment

Perhaps the first thing to notice is that we have been talking about and building models without ever defining just what we mean by the word "model." We even have a section entitled "What is a model?" that never actually answers the question.

This is a deliberate oversight. It is part of our technique for creating an *active* learning environment. Our role is to ask questions. We want you to think about the answers before we tell you how we think about them.

We also want you to *do* things before we discuss them. Notice, for example, how we asked you to criticize models before we even mentioned a basis for criticism. This was just another ploy to involve you, to get you to think actively rather than passively.

Our objective is to stimulate your interest. We want you to *react* to what we write rather than merely absorb it.

In the next few paragraphs we will indeed make an attempt to discuss what we understand by the word "model."

Before you read on, think about what you would have written next. What is a model? Is it a diagram (as in Figure 1.1), or an equation (such as Equation 1.2), or what? How would you describe a model?

We like to imagine you, when you go on to read the next section, talking to yourself. We hope you are saying "No, I am not sure that I agree with that!" or "Yes, I thought of that," or perhaps "Yes, I should have thought of that." The previous paragraph has been indented because it relates, in a sense, to talking to yourself or asking yourself questions. If you page back, you will notice how we have done this before. From now on we will use indented script text whenever we want to draw your attention to how *you* might be thinking. Indented text will be analogous to your conscience—a voice whispering over your shoulder.

Reiterating the Question: What Is a Model?

Figure 1.1 and Equation 1.2 have something in common: they are both *representations* of a book. Figure 1.1 is a descriptive representation. Its purpose is to tell a stranger what to expect from a standard textbook. Equation 1.2 is a mathematical representation, but it also has a purpose. It provides an estimate of the time it will take to read this book, based on the number of pages in the book, the number of exercises, your reading speed, and so on.

We can come very close to a definition of a model by thinking of it as a *purposeful* representation. The word "purposeful" is an essential part of the definition. We cannot build a model if we do not know why we are building it, and we cannot criticize or discuss a model except in terms of its purpose.

It follows that a model does not have to be a *complete* representation. There are exceptions to this. Figure 1.3*a* is a miniature model of a car. For the replica car enthusiast each little detail adds to its authenticity. But this is a special case where completeness and purpose coincide.

In general, a purposeful model includes only those features of reality that are *essential.* Think for example of TV advertisements showing how an automobile manufacturer has used a computer to design a car for minimum air resistance. The advertisement is likely to show a computer screen with a line drawing of the shape of the car such as Figure 1.3*b*. Note how incomplete the line drawing is, but it preserves the one essential feature of the car from an air resistance point of view—its shape.

> *Why do advertisers prefer to use Figure 1.3b rather than Figure 1.3a?*

Presumably because the details in Figure 1.3*a* are distracting. The leaner model makes the point they are trying to make more effectively.

We will learn to admire lean models that capture the essential features of a problem.

There is yet another reason why a model is seldom a complete representation: when we build models we are nearly always constrained by time and resources. Often we want a relatively quick answer (perhaps not a 5- or 10-minute answer, but more likely an answer that we can get in months rather than in years). We want to anticipate what is going to happen rather than wait and see.

After all, if you really wanted the best answer to the question "How long will it take you to work through this book?" you could get it by working through the

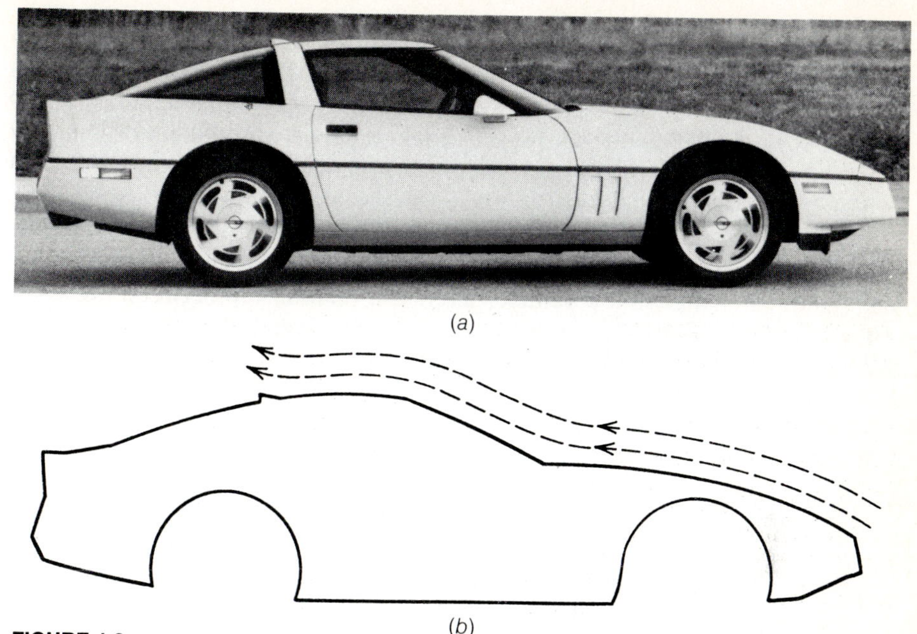

(a)

(b)

FIGURE 1.3
(a) A photo of a modern automobile. This model represents most of the details of the original. *(Courtesy of Chevrolet.)* (b) A line diagram showing only the shape of a car. This representation has a specific purpose: to highlight the air resistance characteristics of the car. It serves that purpose without extraneous detail.

book. If you did that, you would not need a model, only a stopwatch, and the motivation as well as the purpose for building the model would disappear.

SOME MODELING TERMINOLOGY

Variables and Parameters

Equation 1.2 serves as an introduction to some of the terminology of modeling. Remember, the equation is

$$T = pP + wWP$$

where P = number of pages
p = time to read a page
W = average number of tasks per page
w = average time associated with a task

Suppose we were using this formula to calculate the time you should devote not to just this book but to a whole series of books like it. What would change from book to book? Obviously P, W, and w. We therefore call these the *variables* of the model. On the other hand, your reading speed p is unlikely to change from one book to another. We call p a *parameter* of the model.

The distinction between variable and parameter is not always obvious (or even important). It is often a matter of perspective. For example, if we supposed we were using Equation 1.2 to calculate the times that a number of different people would devote to *this one* book, then we would think of p as the variable and P, W, and w as parameters of the model.

Variables may be thought of as those factors in the model that are actively changing. *Parameters* are factors that mediate the effect of the variables on your answer; parameters may be changed too, but in the secondary sense of asking how a change in a parameter alters the relationship between the variable (or variables) and the answer to the problem. Parameters that cannot be changed are called *constants* of the model.

Trade-offs and Sensitivity

You have just received a circular in the mail that offers you the choice between two free courses. The first is a course in speed reading and guarantees that you will be reading technical material twice as fast by the end of the course. The second is a course in what they call "mental gymnastics." It guarantees that by the end of their exercise program, you will be reasoning twice as fast.

Since you are about to embark on the time-consuming task of working your way through this book, you decide it would be well worth your while to take one (but only one) of the free courses.

Which course should you take?

Does the model you built help to answer this question? It might not; after all, it was not designed with this question in mind.

For the purposes of this exercise we suggest you use our model rather than yours. It so happens that Equation 1.2 is a model suited to answering this type of question. Take 15 to 30 minutes to use the equation to help answer the question. You may want to make some calculations, either by using a pocket calculator or a computer. Alternatively, you may be able to argue symbolically directly from the equation.

You are free to make assumptions or invent plausible values for the parameters in the equation.

✔

Remember, Equation 1.2 is

$$T = pP + wWP$$

where P = number of pages
p = time to read a page
W = average number of tasks per page
w = average time associated with a task

Figure 1.4 shows how you might have implemented this equation on a computer spreadsheet. The implementation requires that you provide values for p, P, w, and W. The computer then calculates and displays values for T that are calculated in three different ways:

1. Using the above formula. This is the time it would take you to work through the book without taking either special course.

2. Replacing p by $p/2$ in the above formula. This is how long it should take to work through the book after the speed-reading course.

3. Replacing w by $w/2$. This is how long it should take after the course in mental gymnastics.

Comparing numbers 2 and 3 illustrates the *trade-off* between doubling your reading speed and doubling your reasoning speed.

Once you have implemented this model on a spreadsheet, it is easy to change your estimates for any of the variables or parameters. The spreadsheet then updates the calculations for the three values of T. You could write a small computer program that does this too. Either way, the objective is to make it easy to conduct computer experiments. You want to be able to observe the effect that changes in variables or parameters have on the trade-off between the two courses and hence on your decision to take one course or the other.

There are two interesting conclusions lurking in the structure of the model, waiting, as it were, to be discovered. The first is that the decision is independent of P, the number of pages in the book. The second is a rule: The first course is the one to take when more than half the time spent on the book is reading time.

FIGURE 1.4
The output from a spreadsheet implementation of the reading time model (Equation 1.2).

Time to read a page	Number of pages	Time to perform a task	Number of tasks	Total time
p	P	w	W	T
3	150	10	2	3450
1.5	150	10	2	3225
3	150	5	2	1950

It is unlikely that you would draw these conclusions by idly changing parameters and variables on a spreadsheet. However, the spreadsheet can be used as a laboratory to test hypotheses. You might ask "Does my decision depend on the number of pages in the book?" and then deliberately make changes on the spreadsheet in search of the answer to that question.

Alternatively, you might not have used a computer at all. Instead you might have looked directly at Equation 1.2 and argued as follows. The first course corresponds to halving the first term (pP) on the right-hand side, while the second course corresponds to halving the second term (wWP). It follows that it pays to take the first course whenever

$$pP > wWP$$

or

$$p > wW$$

Notice how this leads directly to the two conclusions drawn above.

How would you present this result? You could state it in words, or you could draw a graph as in Figure 1.5. Here the time to read a page (p) is plotted versus the average time allocated to a task (w). The straight line $p = Ww$ separates the one choice from the other. For points above the line (such as A in the figure), the reading course should be taken, while for points below the line (such as B) the course in mental gymnastics is preferable.

Figure 1.5 is a good way of presenting the result because anybody can plot a point that corresponds to their particular reading speed and reasoning speed and see at once which decision to take.

Notice that the demarcating line in Figure 1.5 depends on W. It is assumed we know the average number of tasks per page. It is important to ask ourselves how *sensitive* our decision is to the estimate of W. For example, what would happen if W were 20 percent larger? or 20 percent smaller?

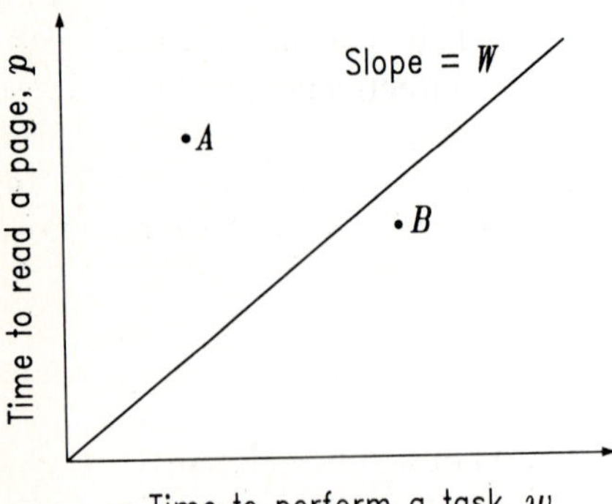

FIGURE 1.5
Graphing the time to read a page (p) versus the time to perform an average task (w) gives a straight line. This helps you decide whether an improvement in your reading or reasoning skills would benefit you more. The answer depends on whether your current skills plot as point A (above the line) or point B (below the line).

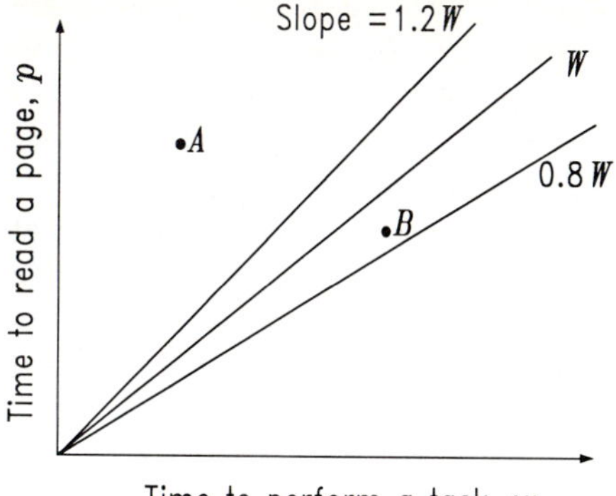

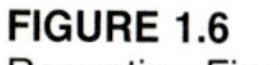

FIGURE 1.6
Repeating Figure 1.5, this time showing the sensitivity of your decision to variations in the average number of tasks per page (the parameter W).

Figure 1.6 shows the three straight lines: $p = Ww$, $p = (1.2W)w$, and $p = (0.8W)w$. It illustrates how our decision might be affected by changes in W. In this case the decision for point B might be reversed if our estimate of W is incorrect, while the decision for point A is more *robust*. It is independent of even quite large changes in W.

Now go back to your model and look at it again. Is there any way you could have argued these points from it?

Resolution, "Lumping," Heuristics, and Algorithms

These are terms that we mention here only to whet your appetite and so you can look out for them. We could use this example to illustrate them (and other words and terms such as stochastic, deterministic, simulation, and strategy) but they all emerge more naturally (and more excitingly) from problems in later chapters.

CONCLUDING REMARKS

It is time we moved on. But before we do so, we hope you have recognized that we *all* use models. Sometimes we build them deliberately, but often we do so subconsciously. In this book we will try to help you build models consciously (or explicitly), and we will emphasize the effectiveness and usefulness of carefully structured, explicit models.

Notice, for example, how the model developed above helps us think and make decisions about reading this book, without ever having to sample more than a few pages.

FURTHER READING

The Search for Solutions by H. F. Judson (Baltimore, Johns Hopkins University Press, 1987) is a book that uses a series of examples and anecdotes to introduce the reader to the methods and accomplishments of science. The chapters on Modeling and Strong Prediction describe the activities of model builders and the process of modeling in a lively and readable manner.

An Introduction to Models in the Social Sciences by C. A. Lave and J. G. March (New York, Harper & Row, 1975) follows a similar format to our book; the reader is asked to stop and think at regular intervals. The problems in the book (problems of demography, learning, behavior, decision making, and individual choice) require careful model building and interpretation.

COMPUTER TOOLS

A computer program called a spreadsheet was introduced in this chapter. A *spreadsheet* is an electronic grid (similar to an accountant's ledger book) that can be used to collect and calculate data. Most spreadsheet programs also have advanced graphics features which let you translate the figures in your spreadsheet into effective visual representations.

Any spreadsheet program will do for representing and solving the problems in this book. If you're familiar with Lotus 1-2-3, Microsoft Excel, Borland Quattro, etc., go ahead and use it. If you're not familiar with these full-capability spreadsheet programs, we suggest that you get a very low cost spreadsheet, such as VP Planner, or the student versions listed below:

1 Canale, R. P., and S. C. Chapra. *Electronic Toolkit: Integrated Computational Software for Engineers and Scientists.* New York: McGraw-Hill, 1988.
2 Sondak, N. *A Guide to Spreadsheets Using VP Planner.* San Francisco: Holden-Day, 1987.
3 *The Student Edition of Lotus 1-2-3.* Reading, MA: Addison-Wesley.
4 *Quattro.* Scotts Valley, CA: Borland International. Full version available at special price for students.

Further information on the use of spreadsheets in engineering, science, and mathematics can be found in:

1 Arganbright, D. E. *Mathematical Applications of Electronic Spreadsheets.* New York: McGraw-Hill, 1985.
2 Orvis, W. J. *1-2-3 for Scientists and Engineers.* New York: Sybex, 1988.

Alternatives to spreadsheets are the equation-solving computer programs, such as:

1 TK Solver Plus (College Edition). New York: McGraw-Hill.
2 Eureka: The Solver. Scotts Valley, CA: Borland International, 1987.
3 MathCAD. Cambridge, MA: MathSoft, Inc.

These are very powerful programs for solving most mathematical functions; however, the solution algorithm is implicit—you do not see how the computer manipulates the model.

CHAPTER 

TIME FOR PING-PONG?

SOLVING A PROBLEM IN 60 SECONDS

Look around the room you are sitting in. Then take just 60 seconds to answer the following question:

"How many ping-pong balls could you fit into the room?"

✔

Time's up! What is your answer?

How did you get it? Did you guess? Did you build a model? If so, can you describe your model?

✔

Sixty seconds is not much time, is it? Perhaps it was just enough time to shrug your shoulders, look around the room, and write down "Lots!" or "Thousands!" These would not have been good answers if we had given you more time, but there is nothing wrong with them here.

What have you accomplished if your answer was "Lots" or "Thousands"? What type of model did you use?

You probably did not think you were using a model at all, but you were! You modeled in terms of categories such as few, some, lots or tens, hundreds, thousands. By drawing a mental picture of a ping-pong ball and looking around the room, you then estimated that the answer belonged in one category rather than another. This is not a useless exercise; it makes a difference whether the answer is "some" or "lots."

Alternatively, you might have tried to make a rough calculation of the number of ping-pong balls. You might, for instance, have estimated how many balls would fit on the wall at one end of the room, and then multiplied by the number that would fit along the length of the room. Or you might have estimated the volume of a ping-pong ball and the volume of the room, and divided the one into the other.

If you made a volumetric calculation, what was your model of a ping-pong ball? What was your model of the room?

We would bet that your model of the ball was a cube rather than a sphere! And you probably modeled the room as a large, empty box. Is Figure 2.1 or Figure 2.2 a fair representation of your model?

What other simplifications or assumptions did you make?

Sixty seconds did not give you much time to think about the assumptions you were making, but you were probably fleetingly aware of some of them. Did you wonder, for instance, about whether you could ignore the furniture in the room? Did you assume that you were not allowed to squash or deform the ping-pong balls?

SOLVING THE SAME PROBLEM IN 5 MINUTES

Now take 5 minutes to solve the same problem. This time keep a note of how you go about solving it. If possible, find a partner to work with.

✔

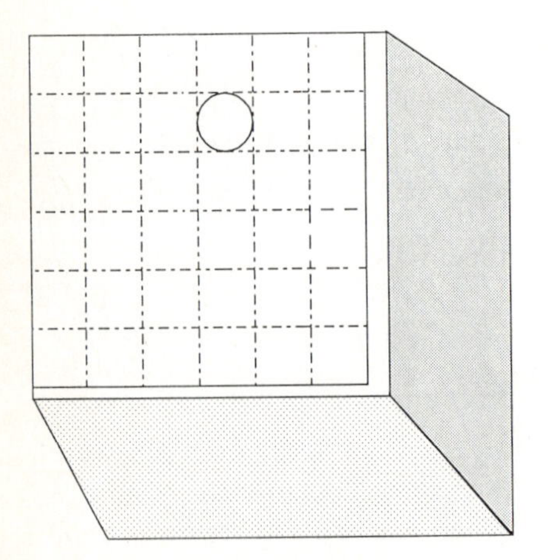

FIGURE 2.1
A 60-second representation of ping-pong balls in a room. (Where is the room? Where are the balls?)

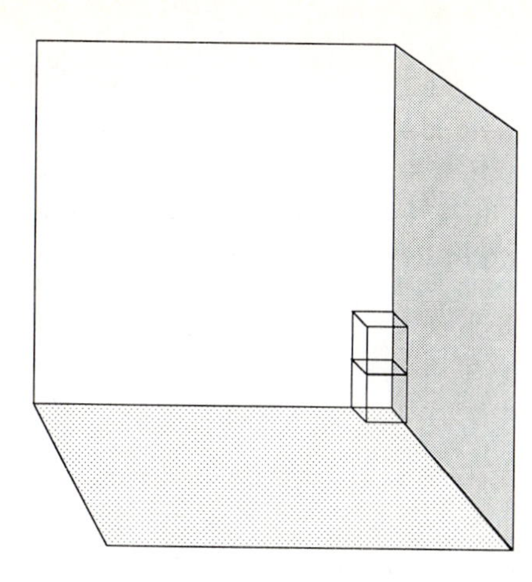

FIGURE 2.2
A 5-minute representation of ping-pong balls in a room.

How did you go about solving the problem this time? Did you use the same procedure but refine your measurements? Or did you use the extra time to take a new approach? Did you change your model?

Did you modify your assumptions? Did you, for example, make a correction for the furniture or the shape of the room?

If you worked with a partner, what were the differences between working together and working alone? Did you share tasks? Did you start with similar ideas, or did you spend time arguing about the proper approach?

Given more resources (time and people), the chances are that you built a more sophisticated model. If your first model was a rough approximation, you might well have switched to a volumetric method such as the model shown in Figure 2.2.

You might even have used a symbolic representation and introduced a notation. For instance, you might have said let

L = length of the room
W = its width
H = its height
D = diameter of a ping-pong ball

Then the volume of the room is

$$V_{\text{room}} = LWH$$

and the volume of a ball (treating it as a cube) is

$$V_{\text{ball}} = D^3$$

and so

$$\begin{aligned}\text{Number of balls} &= V_{\text{room}}/V_{\text{ball}} \\ &= LWH/D^3 \qquad (2.1)\end{aligned}$$

WHAT HAVE WE LEARNED ABOUT MODELING?

Take about 10 minutes to make a list of the points that you think these two exercises illustrate. This is not an easy task, so when you have run out of ideas, read on.

We will first make a list of points that we think are important, so you can compare it with your list. Then we will expand on the more important points.

1. Both exercises illustrate the point made in Chapter 1 that a model is a partial rather than a complete representation.

2. Even a very rough answer is better than no answer at all.

3. A model that is inadequate under one set of circumstances may be the best that you can come up with under another set of circumstances. It follows that the design of a model depends as much on circumstances and constraints (of money, time, data, or personnel) as it does on the problem that is being solved. It also follows that the assumptions one makes depend on the circumstances in which one solves the problem.

4. A symbolic representation (choosing a notation and building a formula or formulae) is "clean" and powerful. It communicates, simply and clearly, what the modeler believes is important, what information is needed and how that information will be used.

5. Sometimes one uses models implicitly (without being aware that one is doing so); at other times one consciously or explicitly constructs or uses a model. An explicit model is an indispensable tool for solving problems and for talking about the solution.

Did you recognize the last point? Would you agree that it is probably the most important point in the list?

We will expand on number 5 after the next section, which relates mainly to numbers 1 and 3 above.

The Real World, the Model World, and Occam's Razor

The room you are sitting in is the *real world* of this problem. It is possible that the room has an odd shape, curved walls or a vaulted ceiling. There are almost certainly windows and doors. One wall may be painted white, another blue. There may be pictures hanging on the walls, carpets on the floor and lights suspended from the ceiling. There is furniture in the room—chairs, desks, cupboards, etc. And for some obscure reason somebody wants to fill it with ping-pong balls.

Your *model world,* on the other hand, is likely to be more like Figure 2.2: a large box filled with small cubes.

What is the connection between these two worlds? How do we get from one to the other? Why, for instance, does the model world have no windows? Does it make sense to ask whether the walls of the box are painted white and blue?

A model can be likened to a caricature. A caricature picks on certain features (a nose, a smile, or a lock of hair) and concentrates on them at the expense of other features. A good caricature is one where these special features have been chosen purposefully and effectively. In the same way a model concentrates on certain features of the real world. Whenever you build a model you have to be selective. You have to identify those aspects of the real world that are relevant and ignore the rest. You have to create a stripped-down model world which enables you to focus, single-mindedly, on the problem you are trying to solve.

William of Occam (or Ockham) was a fourteenth century English philosopher who propounded a heuristic in Latin "*Non sunt multiplicanda entia praeter necessitatem.*" Translated literally, this means "Things should not be multiplied without good reason," but in the context of philosophy (and modeling) it means that one should eliminate all unnecessary information relating to the problem that is being analyzed. Since Occam was reputed to have a sharp, cutting mind, this heuristic is known as Occam's razor.

It is sometimes useful to think of the model world connected to the real world by a tunnel or passageway as in Figure 2.3. The passageway is guarded by a mythical customs officer wielding Occam's razor. It is his job to make sure that nothing inessential is able to pass from the real world to the model world.

One feature distinguishing a good model from a bad model is the way in which Occam's razor has been used. A bad model is the result of either using the razor too little (letting irrelevant details creep into the model) or using it too much (cutting out essential features of the real world). A good model is one that retains a proper balance between what is included and what is excluded.

How do we reach that balance? How does the customs officer do his job? Certain decisions are easy. For example, nobody would even think of asking whether the colors of the walls should be included in the ping-pong model world. But other decisions are not so easy. Should the furniture be in the model world, or should Occam's razor cut it out? And is a cube a good representation of a ping-pong ball?

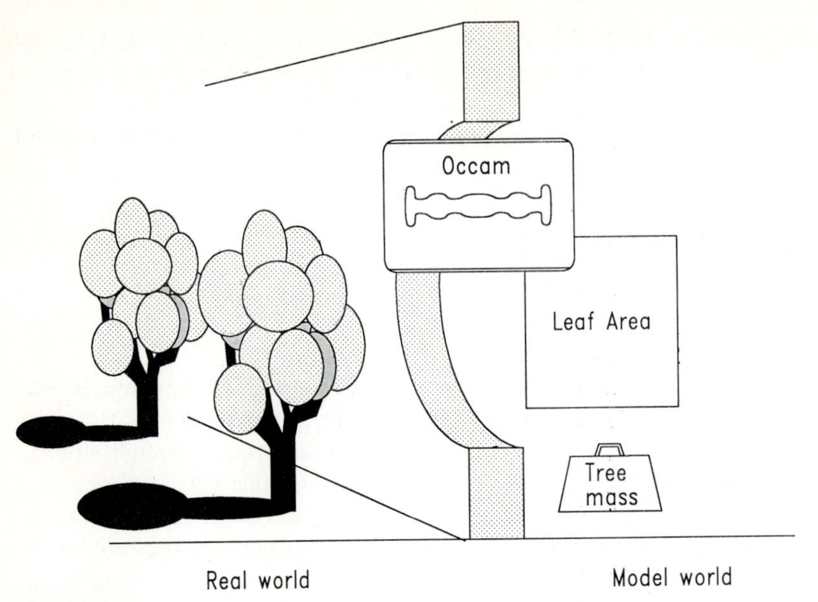

FIGURE 2.3
Occam's razor guards the passage from the real world to the model world. Nothing inessential shall pass by it, but as you abide by its dictum, be careful not to omit the essentials.

There are no hard-and-fast rules. The customs officer does not have a list of forbidden imports. But he does have two important guidelines:

1. *The purpose of the model.* We defined a model in Chapter 1 as a purposeful representation. The customs officer should cut out any aspect of the real world that does not contribute significantly to the stated purpose of the model.

2. *Constraints of time, money, personnel, and information.* The more restrictive these constraints, the more ruthless the use of Occam's razor.

How did we apply these guidelines in the ping-pong problem?

The time constraints were so severe that we almost exclusively invoked the second guideline. Figure 2.1 represents the model we thought of in just 60 seconds. If you think about it, the model world corresponding to that figure is incredibly austere: the room is represented by one wall and a length, while the ping-pong ball is represented by a square cross section or a diameter. Even in the 5-minute exercise, our model world consisted of only a box and cubes, as in Figure 2.2.

Notice that if we had been given more time we would still not have been able to use the first guideline effectively for the simple reason that the purpose of the ping-pong problem has not been stated clearly enough. We do not know whether the object of the exercise is to obtain a rough estimate (to the nearest ten-thousand, say) or whether a significant prize has been offered for the plan that crams in a maximum number of balls without deliberately squashing any of them. In the first case the windows, doors, and light fixtures would be irrelevant, no matter how much time we had for building the model. In the second case, their inclusion or exclusion from the model world would depend only on constraints.

Attempting to Define Heuristics

In Chapter 1 we described a heuristic as a rule of thumb and we flagged it as a modeling term that we would explain more fully later on. We used the word again in the above section; it is time we tried to define it.

A heuristic is a plausible or reasonable approach that has often (but not necessarily always) proved to be useful; it is not guaranteed to be useful or to lead to a solution. This is just a more formal way of saying that it is a rule of thumb. Heuristics are difficult to define, but relatively easy to recognize.

Occam's razor is not the only heuristic introduced in the previous section. What other heuristics can you identify?

The guidelines to the mythical customs officer are heuristics too. They are heuristics for applying Occam's razor.

Notice how the discussion toward the end of the last section illustrates the point that a heuristic need not always be useful. Using the objectives of a model to filter out those aspects of the real world that should be included in the model is a very important heuristic, but it is almost irrelevant when time constraints predominate (as in the 60-second exercise).

It is precisely because we cannot guarantee the efficacy of heuristics that modeling is an art rather than a science. We will have more to say about this in the next chapter.

Why Models Are Important

Have you noticed that we never once said "Build a model"? We just presented you with a problem and then suggested that either consciously or subconsciously you were using models to solve it. We then went on to talk about the models you were probably using.

The point we want to make is that thinking *consciously* and *explicitly* about models is a crucial part of problem solving in general.

Why do you think we have stressed the words consciously and explicitly?

Think about what your answer to the ping-pong ball problem means to the person who asks for it. It does not help very much to give an answer like "lots" or even "28,517 balls" without also giving an explanation of how you got there. The model you have used is as important as your answer in a problem like this.

Why?

Because in the time available there is no way that you are going to solve the problem precisely. You are bound to take shortcuts. It follows that there is no way to evaluate your answer unless one knows more about the assumptions you made and the shortcuts you took. If you are not aware of the model you are using, you are not going to be able to communicate your answer in a meaningful way.

Come to think about it, if you are not aware of the model you are using, you are not going to be able to communicate with yourself. We all need to have an explicit model (or models) to think clearly about any problem we are trying to solve.

This is why models are so important!

It is possible that you noticed the role of models in communication when you teamed up for the 5-minute exercise. Did you ask your partner (or did your partner ask you) "What did you do for the 60-second exercise?" Or did you ask "What was your answer?"

It was far more important to identify your models than to compare answers. If you compared models and then discussed what to do next, you probably did a much better job of the 5-minute exercise than if you jumped in and both argued from different premises.

IF WE HAD MORE TIME

What Is the Best Answer We Could Give?

It is important to consider the best answer you can give to the question, "How many ping-pong balls could fit in this room?" Think about this and discuss it with your partner before giving an answer.

Do you recommend to "measure the room and ball accurately"? Or do you think that the best answer is "Fill the room up with ping-pong balls and count them!" Can you think of a better answer than that? Is it worth the effort? How good an answer are you willing to accept?

Notice that you cannot really answer this question unless the purpose or objective of the exercise is stated more clearly! We are back to the difficulty we noticed when we were discussing guidelines for the use of Occam's razor: Nobody has told us why we are filling the room with ping-pong balls.

But it is always a good heuristic to ask about the best answer one could possibly find!

Thinking about the best possible answer is equivalent to drawing up a wish list of the things you would like to have in your model world. Asking how good an answer you are willing to accept then helps to prune that wish list. This is where the time and other constraints become important.

It is good to ask the question about the best answer even when, as in this case, you find you cannot really answer it. At least you are alerted to the deficiencies in the statement of the problem. In this case you should be prompted to go back and ask "Why do you want to know how many ping-pong balls will fit in this room?"

Upper and Lower Bounds

In the 60-second exercise you were so pressed for time that your objective was to find an answer, almost any answer, without considering how good an answer it was. The 5-minute exercise may have given you more time to be critical about your answer. Certainly, if you had more time it would be prudent to evaluate or put error bounds on your answer.

If you used a volumetric model such as Figure 2.2, did you pause to consider whether your answer would be too small or too large? Assuming that you know the volume of free space in the room, the figure you come up with will underestimate the number of balls.

Can you think of a model that would just as quickly overestimate the answer?

Suppose you model a ping-pong ball as a sphere instead of a cube. If its diameter is D then its volume is

$$V_{ball} = (4\pi/3)(D/2)^3$$

Now imagine the spheres somehow packed into the room so that there are absolutely no air gaps between them. Equation 2.1 then becomes

$$\begin{aligned} \text{Number of balls} &= V_{room}/V_{ball} \\ &= (LWH)/[(4\pi/3)(D/2)^3] \qquad (2.2) \end{aligned}$$

Equation 2.2 gives an upper bound to the answer.

How close are the two bounds?

Notice that if you divide Equation 2.2 by Equation 2.1 you get

$$\text{(Upper bound)/(lower bound)} = 6/\pi$$

which is a ratio of nearly 2.

This would have been a useful calculation to make, even if you only had 5 minutes to solve the problem, because it tells you that the type of model you are using produces answers that could easily be out by 50 percent or more, if only because you have not properly considered how the spheres will fit or pack together.

If you had more time, this calculation becomes even more significant. Should you spend the extra time looking at packing patterns of spheres, or is the volume of furniture or shape of the room more important? Now that you have an estimate of how important the packing could be, you can compare it with, for example, an estimate of the volume of the furniture as a percentage of the volume of the room.

Notice, by the way, how a symbolic or mathematical representation can produce unexpected benefits. Comparing Equations 2.1 and 2.2 enabled us to estimate the ratio of the upper to the lower bound without measuring the dimensions of either a ball or the room.

Comparing Assumptions

Making assumptions is an integral part of deciding what to take from the real to the model world. It follows that assumptions are subject to the two guidelines that help the customs officer to use Occam's razor intelligently. The assumptions one makes depend on both the purpose of the model and the constraints under which it is built.

Asking what we would do if we had more time illustrates how our assumptions depend on constraints. Let us make a list of some of the more important assumptions implicit in Figure 2.2:

- The room is shaped like a box
- A ping-pong ball is assumed to be a cube
- Furniture in the room can be ignored
- Window and door spaces and other nooks and crannies can also be ignored

Given a little extra time, which of these assumptions should we relax? Should we try to estimate the volume of the furniture? Or improve our model of the shape of the room? Or should we investigate the way in which spheres can be packed together?

If we had a lot of extra time we would probably do all three. If we have only a little time we should invest a part of that extra time *ranking* the assumptions. We already (from the section on upper and lower bounds) have some idea of the effect of packing; we should also be able to make rough estimates of the effect of furniture and the shape of the room.

Notice that as we relax the resource constraints, we can afford to consider more and more obscure assumptions. (What about the space under the furniture?) Each assumption we investigate in turn opens up new assumptions at a finer level of detail. Notice too that each time we address an assumption, we make our model world more detailed, and each detail will require more data in the form of measurements (such as the dimensions and shape of each piece of furniture). Eventually we have to establish the objective of the exercise and ask whether the detail is really necessary. In other words, we have to determine the appropriate *resolution* of the model.

Packing Spheres

Suppose we decided that the way in which the balls packed into the room was important. How much difference would that make?

In the section on Upper and Lower Bounds, we considered two extreme cases of how the ping-pong balls might pack into the room. The first was our original model (a ball is a cube) where we ignore the possibility that rows of balls might fit together in a "tight" pattern. In the second model we assumed that the balls packed so tightly that no air space whatsoever was left between them. The two models, respectively, give lower and upper bounds to the number of balls that will fit in a box, and not surprisingly, these bounds are far apart.

To get a better idea of how the balls might actually pack, consider Figure 2.4. In Figure 2.4*a* we ignore packing. This is our ping-pong-ball-is-a-cube model. Notice that the vertical distance between the centers of two balls is just *D*, the diameter of a ball.

Figure 2.4*b* is a more realistic representation. In this case we have denoted the vertical distance between the centers of two balls by *H*.

FIGURE 2.4
A two-dimensional analysis of the extra space made available when ping-pong balls are packed in a tight pattern: (*a*) four balls stacked as squares, and (*b*) tightly packed as circles.

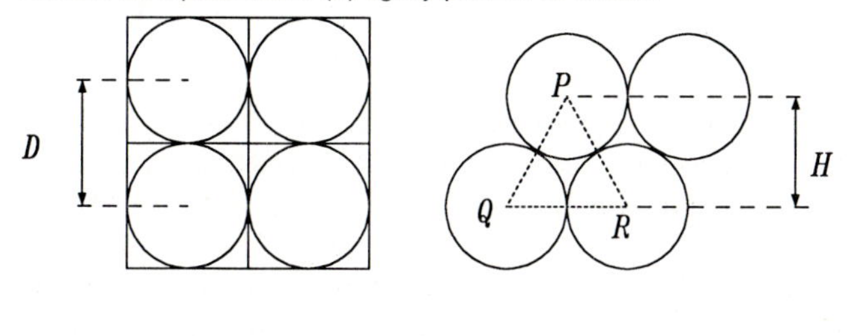

What is the relationship between H and D?

Consider the triangle PQR that joins the centers of three balls. Notice that it is an equilateral triangle and that the length of each side is equal to D. It follows that all three angles of the triangle are equal to 60 degrees and hence that

$$\frac{H}{D} = \sin 60$$

or

$$H = 0.866D$$

Figure 2.4 thus suggests that we could pack nearly 14 percent more balls into the room because of the way they fit.

However, notice that the figure is only a two-dimensional representation of a ball. It is often easier to work with two-dimensional rather than with three-dimensional representations, and it is also often useful to do so. In this case we get a much better estimate of the effect of packing than in either of our previous upper- or lower-bound models. The quick calculation we have made suggests that it would be worthwhile to try to repeat the calculation in three dimensions.

Try to draw Figure 2.4*b* in three dimensions and calculate the vertical distance between the centers of two packed spheres.

Alternatively, can you guess what the answer will be from the two-dimensional figure?

LEARNING TO LEARN

Were the two tasks instructive? How have you benefited from thinking about this problem? What have you learned about "learning to learn" from this exercise?

Compile a list of points about how rather than what you have learned.

✓

The ping-pong problem is one we have used with a large variety of audiences, from junior high school students to professional engineers. It originated (as far as we know) with a professor at the Massachusetts Institute of Technology and came to us by way of Billy Koen (whose book we refer to at the end of Chapter 3). The problem is disarmingly simple, but it helps develop several important skills and illustrates a number of sophisticated modeling concepts:

• It develops the courage to make "back of the envelope" calculations and the wisdom to recognize when they are appropriate

• It also develops the ability to match model resolution with available resources and to be critical of the resolution of a model

• It encourages a pragmatic awareness of assumptions and of the trade-offs between assumptions

• It illustrates the power of symbolic representations

There are several points to be made about the way in which the problem was presented to you and the tasks you were asked to perform:

1. We deliberately provided you with progressively increasing resources. This should have reinforced the points we wanted to make about constraints, assumptions, and resolution. We tried to create an environment in which you would perceive these points before we discussed them.

2. In a classroom we would have

• Asked each pair of students to describe the method they used to solve the problem in the 5-minute exercise

• Recorded the answers and a brief description of the methods on the board

• Asked questions such as "Which is the best answer?" and "Why do you prefer one method to another?" or "What are the strengths of the different methods? And the weaknesses?"

3. Topics such as resolution can then be discussed, pointing out that with more resources (time and people) the answer is better. These discussions would highlight the points we have made in this chapter about resolution. The classroom experience would show up how some students used the same procedure both times but refined their measurements, while others changed methods when more resources were available.

4. Notice that either way (in this book or in the classroom) our approach is to ask you to do something—be it to write, list, construct, or discuss. This forces you to commit yourself to a position before you read or hear what we or the teacher has to say. If you have developed ideas that are similar to ours, you will have learned them far more effectively than if you had read them without thinking about them ahead of time. If your ideas are different, we hope you will either defend them (it could be frustrating trying to argue with an intangible author!) or think about why our ideas are better than yours.

5. Asking you to work in pairs also serves a purpose. The two of you are likely to have somewhat different perspectives or ways of thinking about the problem. Comparing your ideas is likely to help both of you; you are more likely to think actively about the problem if you are working with another person.

6. In this problem there was also a motive in asking you to work alone for the 60-second exercise and then to cooperate on the 5-minute exercise. We wanted to expose you to the need to communicate, to talk explicitly about your model even if you did not realize at the time that you were building a model.

Did you struggle with the assignments? Were you perplexed and unsure of yourself?

We deliberately tried to create an atmosphere in which you were likely to struggle and be perplexed. Struggling is a precursor of learning and being temporarily perplexed is a natural phase in problem solving. You must learn to struggle with that uncomfortable feeling of taking risks in exploring the unknown.

You must also learn to look back on that struggle and evaluate what you have done. George Polya, the famous mathematician, wrote a wonderful little book on problem solving, entitled *How to Solve It: A New Aspect of Mathematical Method.* In it he recommended "looking back" as a heuristic that should come after other heuristics such as "understanding the problem," "devising a plan," and "carrying out the plan." Looking back is a good heuristic for problem solving, but it is a *vital* heuristic for learning.

Finally, have you noticed how we have kept on asking you questions all over the place?

This too has been deliberate. We have been trying to encourage you to think about the problem and also to think about how you are thinking about the problem!

The type of questions we asked are questions you should eventually be asking yourself.

SIMILAR PROBLEMS

You could have a great deal of fun thinking of problems that could be solved using the concepts introduced in this chapter: imagining a "real world" and a "model world," exercising Occam's razor, and observing how your model depends on the time and resources at your disposal. For example: You are about to enter a large room that contains 200 strangers, each wearing a name tag. Estimate how long it will take you to find Mr. Peter Grump. You know he is in the room. First take 60 seconds to make your estimate. Then refine your estimate in 5 minutes, and so on.

FURTHER READING

Comparisons of Distance, Size, Area, Volume, Mass, Weight, Energy, Temperature, Time Speed, and Number throughout the Universe by The Diagram Group (New York, St. Martin's Press, 1980) contains a large number of interesting examples and is, incidentally, superb for assisting in the development of approximation and order-of-magnitude estimation skills.

There are several excellent books that provide morphologies or prescriptions for solving problems. The most famous is probably G. Polya's *How to*

Solve It (Princeton, NJ, Princeton University Press, 1957), which we mentioned in the previous section. Polya's book is the inspiration for much of our thinking (and also for the title of this book).

You might also enjoy reading *Conceptual Blockbusting* (New York, Freeman, 1976) by J. L. Adams or *The Ideal Problem Solver* by J. D. Bransford and B. S. Stein (New York, Freeman, 1984).

PURGING A GAS STORAGE TANK

THE PROBLEM

A gas storage tank has a volume of 3000 m^3. It currently contains methane. The tank must be emptied so that it can be cleaned and inspected. Safety regulations require that it should contain no more than 1 part in 100 of methane before work (which may include welding) can be started.

Nitrogen is available and can be pumped into an opening near one end of the tank. Another opening (near the other end) will let gases escape. How much nitrogen will you need to dilute the methane effectively?

YOUR FIRST TASK

Your first task is to develop an approach to this problem: in other words, a plan for determining how much nitrogen will be needed.

Take about 30 minutes to prepare your plan and to come up with a representation that helps to explain it to others. Your representation could be a diagram, a set of equations, a flowchart, an algorithm, or whatever format you choose.

If you find it difficult to develop a representation, then make a list of questions you would like to ask, or assumptions you would like to make.

Do not actually try to solve the problem at this time. Also, remember that the decision to use nitrogen has already been taken; you may have thought of using water to purge the methane, or have had other bright ideas, but that is not the problem you have been asked to solve.

✔

WHAT HEURISTICS HAVE YOU USED?

While your plan is still fresh in your mind, think about how you actually developed it. What heuristics did you use? What ideas did you have?

When you have thought about what you did, read on.

✔

HEURISTICS YOU MIGHT HAVE USED

Quickly read the following list, and note the heuristics you used in formulating your plan of action. Did you:

Rephrase the problem to make sure you understood it?
Draw a simple diagram of what was happening?
Think about what was going into the tank and what was coming out?
Imagine yourself inside the tank, and ask what was going on around you?
Ask whether there were any physical laws to consider (such as conservation of matter or energy)?
Try to imagine the answer as a number, graph, table, or whatever?
Try to identify essential variables?
Choose a notation?
Look for a ready-made formula for the answer?
Look for simplifying assumptions?
Try to find an easier version of the problem?
Look for bounds (simple models that would definitely underestimate or overestimate the answer)?

WHERE DID YOUR HEURISTICS LEAD YOU?

We would like to look at what you have done and discuss it with you. Unfortunately, we cannot do that. In this section we will do the next best thing. We will present ideas that students have had when we have used this problem in class, and we will comment (and encourage you to comment) on their ideas. While we do this, we want you to keep in mind what you have done. In this way you can compare their ideas with your own.

We are not going to present *all* the ideas that students have had, because we do not want to take all the mystery out of the problem. We want you to concentrate on your solution rather than theirs or ours.

Questions Students Asked or Assumptions They Made

Many asked "What is the pressure of the methane in the tank?" Some students answered this for themselves. They guessed that the pressure in the tank was likely to be higher than atmospheric pressure. It would then be sensible, they argued, to release all excess methane before doing anything else. They therefore assumed that nitrogen would be added to a tank that was initially full of methane at atmospheric pressure.

"Do nitrogen and methane react chemically?"

If you have not studied chemistry this is an important question to ask. The answer is "No."

"Is methane denser than nitrogen, or vice versa?" Nitrogen is denser.

"So will the nitrogen push out the methane, or will the two gases mix in the tank?"

"If they do mix, will it be a thorough mixing or will there be layers of different concentrations in the tank?"

These are both good questions. If you cannot answer them from your own knowledge or experience, then you should ask a physicist or chemist to help you make a good assumption. Either will tell you that mixing will definitely take place. It is better to assume that the gases mix thoroughly than to assume that they do not mix at all.

Simple Diagrams and Simple Models

Most students started by trying to simplify the problem.

Figure 3.1 is a diagram one student drew to represent a simplified model. It is one that ignores the advice from the physicists and chemists. It assumes that the nitrogen and methane cannot mix. The student imagines a membrane between the two gases. As nitrogen enters the tank it pushes the membrane up and forces pure methane out.

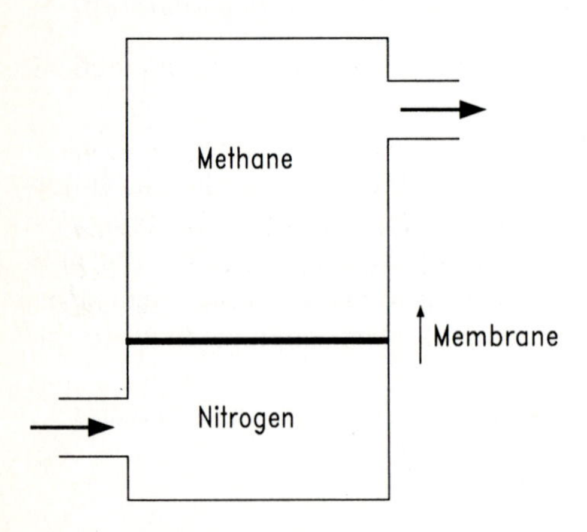

FIGURE 3.1
A representation that assumes no mixing can occur. As nitrogen enters the tank, an imaginary membrane separating the gases pushes out the methane.

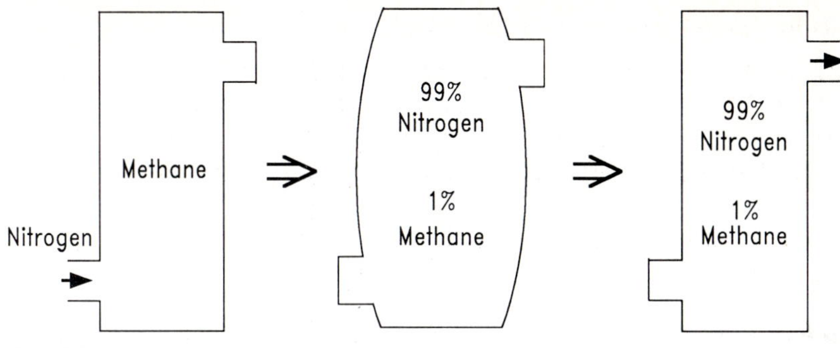

FIGURE 3.2
The "bursting" model. Nitrogen is pumped in, but nothing is let out until the methane concentration has dropped to 1%.

Figure 3.2 is another student's representation of a simplified model. This student assumes that nitrogen is pumped into the tank without letting anything out. Only when the mix inside the tank is 99% nitrogen is the mixture released.

Which model do you prefer? Why?

✓

We like both models.

"But," we can almost hear you say, "The first model is totally unrealistic! We know that there is no membrane between the two gases, and we have been told that the gases will mix. The nitrogen is not going to push out pure methane!"

"The second model is even worse. It is totally impractical. It would require such high pressures that the tank would rupture or even explode!"

We like both models *despite* the fact that they are unrealistic and impractical.

Why do you think we like them? And why do you think the students thought of them?

See if you can guess at the heuristics these students might have used. In particular, think about how they would have reacted to the list of heuristics presented in the previous section. Can you find any parallels between their approaches and your approach to this problem?

Think about these questions before you read on.

✓

We suspect that both students had trouble thinking about how the two gases mix in the tank. So they tried to get a handle on the problem by simplifying it.

The first model (Figure 3.1) is really a "thought experiment." It asks "What kind of answer would we get if the gases do not mix?" The second model (Figure 3.2) emerges from questions like "Is it possible to create a condition in the tank that would enable us to *know* the composition of the mixture that is released?"

Both models avoid the difficulty that in reality the composition of the gas leaving the tank changes during the process of pumping in the nitrogen.

Did you recognize this as an important consideration?

It seems likely, therefore, that both students used the heuristic "Try to find an easier version of the problem; one that you might be able to solve." They probably realized that they could follow these models through and get answers out of them. After all, it is better to do this than nothing at all. They might even have hoped that working with a simplified model would trigger ideas on how to improve it.

That is why we like both models and why we have presented them here. We like to think of the two models as "thought experiments" which help us to come to grips with the problem. We hope that they will give us clues that lead to better models.

The students might have reached these representations via a different heuristic. They may have tried, deliberately, to find an upper or lower bound to the answer. Since the gas leaving the tank will be a changing mixture of nitrogen and methane, a model where no nitrogen can leave must *underestimate* the amount of nitrogen that has to be pumped in. The representation in Figure 3.1 therefore provides a lower bound.

The representation in Figure 3.2 overestimates the amount of nitrogen and provides an upper bound. Why?

Some students used the lower-bound model, others the upper-bound model. Very few used both in their first exercise.

What are the advantages of having both models? What question or questions would you want to ask next?

✓

An important question at this point is "How close are the two bounds?" If they are close, there is no need to develop more sophisticated models.

If you have not already done so, go ahead and calculate the numerical answers for the models in Figures 3.1 and 3.2. Do this before doing anything else.

✓

Upper and Lower Bounds

The lower bound is 2970 m^3, or 1 percent under 3000 m^3, whereas the upper bound is 297,000 m^3.

(Have you checked to see that these answers both reduce the concentration of methane to 1% or less?)

We do indeed have upper and lower bounds, but they are so far apart that they are not particularly useful. We are out of luck!

Which bound do you think is more realistic? Why do you think so? Can you think of ways of improving either of the two models? If so, write them down.

✓

One student thought of a more realistic version of the upper-bound model. Instead of pumping nitrogen into the tank (without releasing any gas) until the mixture contains only 1% of methane, the student asked what would happen if we only doubled the amount of gas in the tank?

The concentration of methane would be halved.

Suppose we released the excess gas and then doubled the amount in the tank again?

Each time we do this, we pump in 3000 m^3 of pure nitrogen, and subsequently release 3000 m^3 of a mixture of gases. Each time the concentration of methane in the tank is halved.

How many times would we have to do this before the concentration of methane was less than 1%?

The result of the calculation is seven. Seven times 3000 is 21,000 so we would need 21,000 m^3 of nitrogen.

This is still an upper bound, isn't it? Notice how we are closing the gap between our upper and lower bounds.

Notice too that this new upper-bound model is the same basic idea as the old one, but opens up exciting new possibilities. Jumping from the original upper-bound model to this model was the sort of brilliant idea that we call a "cognitive leap."

What are the heuristics that the student might have used (either consciously or subconsciously) to make that leap?

What plan of action do you think this student developed for solving the problem?

✓

YOUR NEXT TASK

What we have seen so far is useful, but we still have not solved the problem. The time has come for you to build a more accurate model.

Decide whether you still like your original plan of action or whether you want to modify it, or even replace it, in view of some of the ideas that have been presented.

Whatever approach you decide to take, your next task is to

- Build a model
- Implement it
- Make sure your results are consistent with your understanding of the model and the problem
- Think about how you would describe your model and results in a formal presentation (in other words, make sure you understand what you have done well enough to explain it clearly)

You may use whatever computational tools you need. You may want to use a pocket calculator or a spreadsheet, or you may want to write your own computer program. You might even be able to solve the problem using only pencil and paper!

Depending on how eager you are, you could spend an evening (or even longer) on this task.

✔

OUR FIRST MODEL

There are many different ways of tackling this problem. Our first model is an extension of the upper-bound model.

In the simplest version of the upper bound model we pumped nitrogen into the tank (without letting anything out) until the concentration of methane was 1%. In the second version we repeatedly pumped in nitrogen, but each time we pumped in only enough to halve the concentration of methane before releasing the excess.

What are the key features of both these versions?

They are

- A tank of fixed volume
- We pump in some nitrogen but let nothing out of the tank
- This reduces the concentration of methane
- We then open a valve which releases the excess amount of well-mixed gases
- We repeat this until the concentration of methane in the tank is less than 1%

How could we improve the accuracy of this answer?

Notice that we improved our estimate of the upper bound from 297,000 m^3 to 21,000 m^3 by breaking the process down from one big step to seven smaller steps. Perhaps we should break it down into even smaller steps. That means we should reduce the amount of nitrogen we pump in each time. The following notation helps describe what we are trying to do:

1. Let the volume of the tank be V. (The model will be more general and easier to manipulate if we think of V rather than an actual number like 3000 m^3.)

2. Assume that each time we pump in only a small amount of nitrogen x.

3. Suppose that this changes the concentration of methane by a factor q. (This factor must be less than 1.0 since we are reducing the concentration of methane.)

If we choose a value for x, can we predict the value of q?

This is the crux of our model. Let us try to develop a relationship between x and q at the same arbitrary time during the purging process:

Suppose that at that time there are v m^3 of methane left in the tank; the rest of the volume is nitrogen. It follows that the concentration of methane is

$$\frac{v}{V}$$

We then pump in an extra x m^3 of nitrogen. Since the amount of methane does not change, the new concentration of methane in the tank is

$$\frac{v}{V + x}$$

Since q is the ratio of the new concentration to the old,

$$\begin{aligned} q &= \left(\frac{v}{V + x}\right)\left(\frac{V}{v}\right) \\ &= \frac{V}{V + x} \end{aligned} \tag{3.1}$$

(Notice that when $x = V$, this gives $q = 1/2$, which ties in with the student's modified upper-bound model. Making checks like this is a good habit to cultivate.)

What is the situation after N such steps? How much nitrogen will we have pumped in? How about

$$Nx \tag{3.2}$$

And what is the concentration of methane in the tank?

If we start with a concentration of 1.0 (all methane) and multiply it by a factor q at each step, after N steps it will be

$$q^N = \left(\frac{V}{V + x}\right)^N \tag{3.3}$$

If we choose a small value for x, it remains to find N. We leave this to you to do. You may want to use a computer, or you may just need a pocket calculator.

✓

We decided to solve this problem on a computer. Figure 3.3 shows our three columns of output:

- The first shows the number of steps
- The second shows the total amount of nitrogen used so far (the number of steps times x)
- The third shows the current methane concentration, starting at 1.0 and multiplying by the factor q (calculated from Equation 3.1) at each step

We manipulated x and checked down the third column to see when the methane concentration dropped below 1%. We then plotted the total amount

FIGURE 3.3
A computer solution for the upper-bound gas tank model. Notice that the methane concentration drops below 0.01 at the 72d step in this particular calculation. The total amount of nitrogen pumped into the tank is 14,400 m^3.

Step Number	Total nitrogen so far, m^3	Concentration of methane in tank
0	0	1.0000
1	200	0.9375
2	400	0.8789
3	600	0.8240
4	800	0.7725
5	1,000	0.7242
6	1,200	0.6789
67	13,400	0.0132
68	13,600	0.0124
69	13,800	0.0116
70	14,000	0.0109
71	14,200	0.0102
72	14,400	0.0096

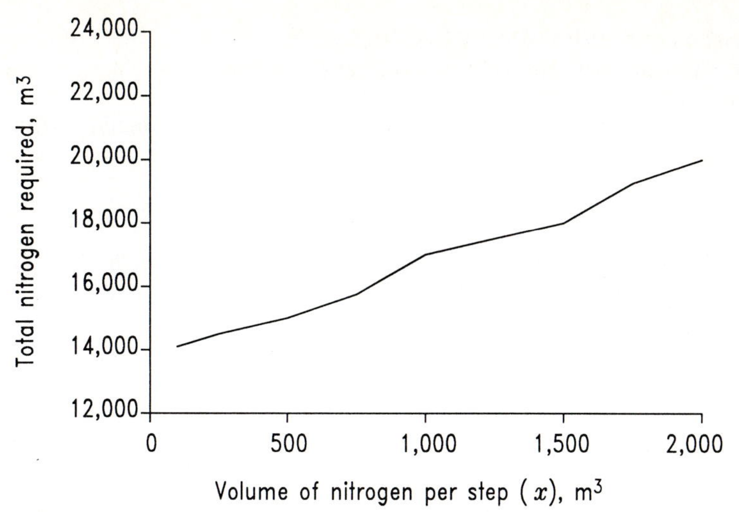

FIGURE 3.4
Results from the upper-bound model, showing how the answer depends on the small amount of nitrogen x pumped in at each step of the model.

of nitrogen (read from the corresponding position in the second column) versus the value of x. Figure 3.4 shows our results.

How much nitrogen do we need? Did your model give a similar answer?

OUR SECOND MODEL

As an alternative, suppose we try to improve the lower-bound model.

How could we do this? Remember, in the lower-bound model the gases do not mix; the nitrogen we pump in pushes out an equivalent amount of pure methane.
What hints could we take from the previous model?

How about pumping in a small amount of nitrogen one step at a time?

Then what will come out of the tank?

The following algorithm preserves the spirit of the lower-bound model:

1. Suppose that at each step we pump in p m^3 of nitrogen. It is useful to imagine it wrapped in an impermeable membrane that separates this small amount of nitrogen from the rest of the gas in the tank.
2. We then release the excess p m^3 of the rest of the gas.
3. We puncture the membrane and mix thoroughly.
4. We repeat the above steps.

The first time we do this, we will release p m^3 of pure methane. How much methane will we release the second time?

There are $(V - p)$ m^3 of methane left in the tank, and so the concentration of methane will be

$$(V - p)/V$$

This is the *proportion* of methane in the tank. Since we have assumed that the gas is well mixed, we expect to find exactly the same proportion in the gas that is released during the second step. It follows that we will release

$$p(V - p)/V$$

m^3 of methane the second time.

How can we generalize this? What is a convenient notation?

The amount of methane in the tank changes from one step to the next. We therefore want a notation that makes it easy for us to identify which step we are talking about. If we let y represent the amount of methane in the tank, then tagging a subscript onto y leads to the notation y_0 for the amount of methane at the start, y_1 for the amount after the first step, and, in general, y_n to represent the amount of methane in the tank after n steps. We call y_n a *subscripted variable*.

The concentration of methane after n steps will then be y_n/V, and so by the same arguments that we used above, the amount of methane released during the next (the $n + 1$) step will be py_n/V. It follows that

$$y_{n+1} = y_n - \text{methane released during step } n+1$$

or

$$y_{n+1} = y_n - py_n/V \tag{3.4}$$

Starting with $y_0 = V$, we can apply Equation 3.4, step by step, until y_n drops below $V/100$. Suppose we find that this takes M steps. Then we will have used Mp m^3 of nitrogen.

What will happen to the answer as we make p smaller? Notice that it is going to take more and more calculations to find M. Notice too that however small we choose p, our answer will always be a lower bound. Right?

This model was also implemented on a computer. In Figure 3.5 we have plotted the final amount of nitrogen (when the methane concentration dropped below 1%) against our choice of p.

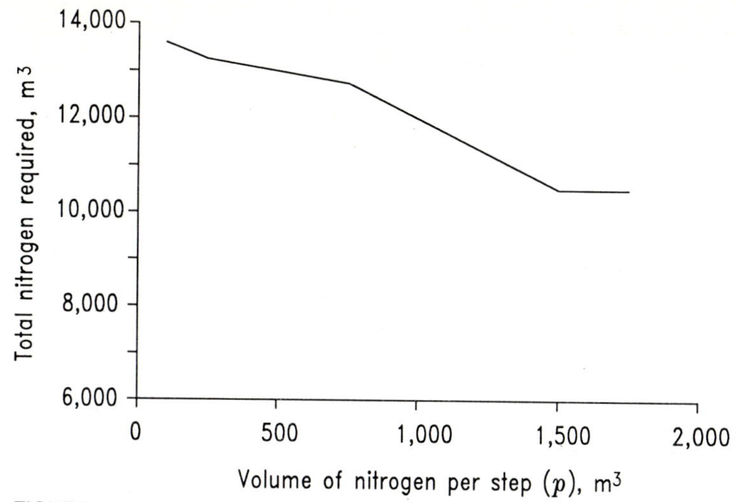

FIGURE 3.5
Results from the lower-bound model (Equation 3.4). It shows total nitrogen required plotted against the amount of nitrogen p added at each step of the model.

THE DIFFERENCE BETWEEN AN ALGORITHM AND A HEURISTIC

Both our models led to a set of instructions. Following the steps of these instructions led to an answer to the problem. An *algorithm* is an unambiguous set of instructions that leads to an answer for a problem.

What is the difference between an algorithm and a heuristic?

The algorithm *guarantees* an answer. It is a recipe rather than a rule of thumb. It is a recipe that is so clear that it can be implemented, for example, in a computer program. The answer it reaches might not be as accurate as you may need, but it will always reach an answer and usually there are ways of improving the accuracy. For example, we could improve the accuracy in our second model by making p smaller.

A heuristic, as mentioned in Chapter 2, is difficult to define but relatively easy to identify using the characteristics listed by Koen in his book *Discussion of the Method* discussed at the end of the chapter. Heuristics help solve unsolvable problems or reduce the time needed to find a satisfactory solution. Their use does not guarantee a solution; two heuristics may contradict or give different answers to the same question and still be helpful. The heuristic depends on the immediate context instead of using absolute truth as a standard of validity.

An algorithm is an attractive way of developing the solution to a problem on a computer. A new heuristic is "Try to develop an algorithm." In other words "Try to develop a series of steps that will lead to the answer you are looking for."

MORE AND LESS USEFUL APPROACHES

We have now seen several approaches to this problem. We have also given some thought to heuristics and how they help to solve the problem or point toward useful approaches.

Heuristics could also be misleading. How do we recognize when a heuristic is pointing us in a useful direction? How do we know which heuristics to invoke?

Judging the usefulness and appropriateness of heuristics brings us to a discussion of the idea of control in problem solving. *Control* is the metacognitive (thinking about thinking) process of asking (and answering) questions such as those posed above.

Unfortunately, there are no easy or general answers to these questions. Control is ideally learned in an apprenticeship. We would like to examine your solution and ask you to describe how you got there. Then we could discuss your control heuristics and possibly offer advice.

Instead, we have to somehow help you to assess and guide yourself. We refer to the book *Mathematical Problem Solving* by Alan Schoenfeld at the end of the chapter. He suggests that you ask yourself the following three questions to promote metacognition:

1. What (exactly) are you doing? (Can you describe it precisely?)
2. Why are you doing it? (How does it fit into the solution?)
3. How will it help? (What will you do with the outcome when you obtain it?)

Bearing these questions in mind, let us look back both on the less and the more useful approaches to this problem and to the heuristics associated with those approaches.

A Less Useful Approach

This problem is deceptive, since at first glance it looks simple. We want to calculate a quantity of nitrogen. It is tempting to call that quantity X and to look for or try to construct a formula to find X. In other words, we might set out to write down an equation of the form

$$X = \text{some expression}$$

or better still

$$X = \text{a function of (what?)}$$

If we followed this approach, the next step would be to ask what the function depends on. Two things immediately come to mind: we expect it to depend on the volume of the tank (V) and on the final concentration of methane

(1%). A little thought suggests that the function must also depend on the way in which the nitrogen mixes with and displaces the methane, but it is far from obvious how to take this into account.

How useful is this line of thought? What exactly are we doing here? (Remember Schoenfeld's questions?)
What heuristics have we used?

We have jumped directly into the middle of our problem by choosing a variable (X) without thinking carefully about our choice, and we have invoked the heuristic "Is there a ready-made formula?" We have not talked about or even thought about a model. And yet the notation we have chosen subtly *presumes* a model! Writing X for the amount of nitrogen encourages us to think of X as a *single package* of nitrogen. This package goes into the tank, something comes out, and we are left with less than 1% methane in the tank.

This is a shoddy approach. Instead of consciously and explicitly developing a model that leads to a suitable choice of variables, our choice of variable has subconsciously or implicitly led to a model. We are not in complete control of what we are doing!

Is it wrong to identify variables and look for formulae?
Probably not. We suspect that you use this approach, with varying success, in your algebra and physics quizzes! So, under what conditions is this an appropriate heuristic?

Looking for a formula can be a powerful heuristic or an intellectual gamble depending on when and how the heuristic is applied. The same can be said of identifying variables. We have gone wrong here because we invoked these heuristics too soon. We should have started by developing our model. The secret is to look for a formula only when you are sure of what you are trying to do. Even then, you should be ready to abandon the search and try another approach if the search for a formula does not look promising.

Having recognized the weaknesses of this approach, it does not necessarily follow that the approach will always fail. Our choice of variable has implicitly led us to a model. If we fail to recognize that this is the model we have unwittingly built, then it is unlikely that we will make any progress. However, if we do recognize the model behind our approach, then perhaps all is not lost. We might, for instance, ask the question "If a package of nitrogen goes into the tank, what comes out of it?" Can you see that this could lead up to either the model represented in Figure 3.1 (we assume that pure methane comes out) or the model represented in Figure 3.2?

Asking "What comes out of the tank?" could also lead to the answer that what comes out is a mixture of gases and the realization that the composition of the mixture changes during the process of purging the tank. Asking how it changes might lead us to ask whether we need a different type of model.

For example, having recognized that there is a difficulty in writing down a formula for X because the composition of the mixture going out is changing, we might have asked "Can we construct X? Can we build it up out of little pieces?" That question could have led us to a solution of the problem.

More Useful Approaches

Thinking about the nitrogen going into the tank can be misleading. It is better to concentrate on the changing mix of gases coming out of the tank. Are there heuristics that could have led us directly to this view of the problem?

Drawing a simple diagram helps us to see the problem in this light. What to draw? Start with a tank with a fixed volume of 3000 m^3. What else? Well, nitrogen is going into it, and something must be coming out. What exactly is coming out? That depends on what is happening inside the tank. We suspect the two gases are mixing, so let us *assume* they are mixing. Therefore a mixture of methane and nitrogen is coming out. A drawing such as Figure 3.6*a* begins to take shape.

Both the restatement of the problem and the drawing lead us into asking what the relationship is between the nitrogen going in and the gases coming out. The conservation principle then tells us that there can be no accumulation of gas in the tank; the volume of nitrogen going in must equal the volume of mixture coming out. How to represent that? Perhaps as in Figure 3.6*b*, by drawing two little balloons of the same size, one for the nitrogen going in and the other for the mixture coming out.

Notice how evocative Figure 3.6*b* is! Just drawing those two balloons suggests a train of thought that could lead to either of our two models. All we have to do is think of a small quantity of nitrogen going in, the same quantity of mixed gases coming out, another small quantity of nitrogen going in, a slightly different mix coming out, and so on. We are irresistibly led to the idea of building up the solution one step at a time.

FIGURE 3.6
A simple representation of the purging process. These diagrams illustrated (*a*) perfect mixing and (*b*) the idea of *incremental* inputs and outputs.

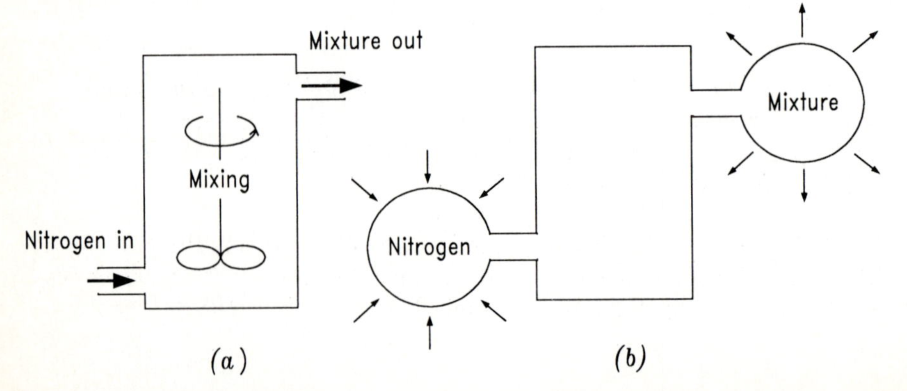

Other heuristics in our list could have had the same effect. For example, thinking of the solution as a table or graph is more likely to lead to a piecemeal approach (constructing an answer step by step) than thinking of the answer as a single number. This is why it so often pays to think of an answer in different, less obvious ways!

As we have seen, searching for simplified versions of the problem, or for upper and lower bounds, can also lead to the solution, albeit in two stages. The first stage is to find a simple model or a bound. The second stage is to ask how we can *refine* that model or bound.

Finally, there is a heuristic that was not on our list that would have been really helpful. It is the salami tactics heuristic. "Salami tactics" is a phrase borrowed from the world of bureaucrats and politicians. They use it in the sense of "If you cannot get the whole salami all at once, try to get it slice by slice." We use it in the sense of "Have you tried to develop a sequence of steps that would lead to the required answer?" In this problem the salami tactics heuristic slices the problem into bite-size chunks and leads to at least two effective algorithms. It is in fact a powerful heuristic because it often pays to analyze a process as a series of small steps.

DISCUSSION

Powerful Heuristics

What have we learned about choosing and using heuristics? What are some of the key features of heuristics?

The heuristics we have considered all share a common feature: they are amazingly versatile.

You might have noticed that it did not matter all that much which heuristics we invoked; most of them had the potential to point us toward a method of solution or at least a constructive way of thinking about the problem.

This versatility seems to be the hallmark of a powerful heuristic. Powerful heuristics generate useful ways of thinking. It follows that it is not all that important to ask how we choose heuristics. It seems that almost any powerful heuristic is a good choice if we use it intelligently.

This raises new questions. How do we know whether we are using a heuristic intelligently? How do we recognize that a heuristic is pointing us in a useful direction? If almost any heuristic will do, why bother to talk about heuristics at all?

The last question is the easiest to answer. We solve problems by *thinking*. It seems as though people need stimuli to provoke thought or to make their cognitive leaps. Heuristics are powerful stimuli.

To change the analogy, heuristics give us the scaffolding that we need to help us to build our thoughts. It does not matter much whether the scaffolding is made out of wood, steel, or whatever. The important point is that the scaffolding should be there: we need it!

Asking how we recognize that a heuristic is pointing in a useful direction or how we know that we are using heuristics intelligently are both much more difficult questions. There are no easy answers and there are certainly no short answers. If we could answer these questions in a paragraph or two, there would be no need for this book.

These are questions about control and we learn control strategies by experience. We learn them faster if we criticize and analyze that experience, if we ask questions like those suggested by Schoenfeld. A good apprenticeship provides both the experience and the analysis. That is what we are trying to accomplish in this book.

IMPLEMENTING THE SOLUTION

How did you implement your solution?

We have talked a lot in this chapter about heuristics and modeling approaches, but very little about actually implementing a model after one has developed a promising approach. You might have guessed from Figure 3.3 that we used a spreadsheet.

Was this a sensible mode of implementation?

From a pragmatic point of view the answer must be "yes"; it was a quick and reliable way to get answers. Moreover both our models lent themselves to implementation on a spreadsheet. There is, however, one major disadvantage to this approach. Both models improve in accuracy as we increase the number of steps in the algorithm. To get the most accurate results may require more columns than are available on a spreadsheet. One alternative is to write a computer program. The other is to look more closely at the mathematics of the two models.

Are there shorter routes to these answers?

Let us look at the mathematics of our first model.

After N steps of the model, from Equations 3.2 and 3.3 we know that we will have used up Nx m^3 of nitrogen. At that stage the concentration of methane in the tank will be q^N or $[V/(V + x)]^N$.

We want the concentration to be less than one in a hundred. It follows that

$$\left(\frac{V}{V + x}\right)^N \leq 0.01$$

or

$$\left(\frac{V + x}{V}\right)^N \geq 100$$

If we take natural logarithms, this gives

$$N[\ln(V + x) - \ln(V)] \geq \ln(100)$$

or

$$N \geq \frac{\ln(100)}{\ln(V + x) - \ln(V)}$$

If we choose a small value for x, we can calculate

$$\frac{\ln(100)}{\ln(V + x) - \ln(V)}$$

and round up to the next integer to find N.

(Why round up, not down?)

We could even be more imaginative and write $N = \ln(100)/[\ln(V + x) - \ln(V)]$, without worrying whether N is an integer or not.

(Why do you think we can do this?)

It follows that the total amount of nitrogen we need is

$$Nx = \frac{x \ln(100)}{\ln(V + x) - \ln(V)} \tag{3.5}$$

We have implemented the model without using a computer at all!

If you have had a course in calculus, how can you improve on Equation 3.5? What happens as x tends to zero?

Notice that the limit, as x tends to zero, of the expression

$$\frac{\ln(V + x) - \ln(V)}{x}$$

is in fact the *derivative* of $\ln(x)$ evaluated at $x = V$.

How does this simplify the solution?

We leave that for you to explore!

It is also possible to implement the second model without a computer. We know that

$$y_0 = V$$

It follows from Equation 3.4 that

$$y_1 = V(1 - p/V)$$
$$y_2 = V(1 - p/V)(1 - p/V)$$

and so on.

We leave it to you to find the pattern and deduce an expression for y_M. Alternatively, if you have taken the appropriate calculus courses, you could try to rewrite Equation 3.4 as a differential equation. Remember, p represents a *small* amount of nitrogen!

Choice of Variables and Notation

In describing a less useful approach, we showed how an unfortunate choice of variables or notation could lead one astray during the formulation of a model.

The converse is also true. Notice how vague a model is until meaningful variables and notation are introduced. Once the notation is there, it suddenly becomes easier to describe and think about a model. It even becomes possible, as we have just seen, to implement a model without even going to the computer!

There are two points to note about the notation we have used:

1. In modeling it is usually better to use symbols rather than numbers. For example, it is better to call the volume of the tank V rather than 3000 m^3. Notice how much easier it was to write and manipulate equations once we started using V instead of 3000 m^3. This is a small but very useful point to remember. Whenever you are developing formulae or algorithms, delay using actual numbers until the very end.

2. A subscript notation (such as y_n in Equation 3.4) is often appropriate when we think in terms of a sequence of numbers or computations. In the first place, the subscript emphasizes that we are dealing with a sequence and helps us to keep our thinking straight. In the second place, writing down an equation such as Equation 3.4 for one step in the sequence often summarizes the whole process. If we know how to take that step, we know how to solve the problem.

Which Was the Better Model?

To answer this question we need to consider what makes one model better than another. There are two important criteria:

1. We first ask "Which model better meets the objectives of the modeling exercise?" To answer this we need to compare the two models to see what they ignore, what they include, what assumptions they make, and how they are structured. The judgments we make here are all related to the *purpose* of the model.

In terms of this criterion, our models are equally good, even though they started from very different assumptions. Recall that the original lower-bound model assumed that no nitrogen was released from the tank, whereas the original upper-bound model assumed that the mixture released was nearly all nitrogen; it contained only 1% methane. However, once we have modified the algorithm so that we make assumptions about only one small step at a time, the two models are almost identical. In each we pump in a small amount of nitrogen at a time. If you think about it carefully, the only real difference is that in the first model we mix in the new nitrogen before we release the excess gas. In the second model, the excess gas is released before mixing. This explains why the first model provides an upper bound (it is as though some of the nitrogen we are currently pumping in goes straight out of the tank) and the second a lower bound. There is no way to argue that the one approach is better than the other.

2. The second criterion relates to the actual algorithm or mathematical formulation of the model. It is a purely technical consideration. We ask whether one model is more easily implemented than another or more robust than another (in the sense that it is less likely to be plagued by computational errors).

In this example, both models lead to similar mathematical expressions or spreadsheet calculations. Again, there is no basis for saying that one is better than the other.

It is always comforting to have upper and lower bounds to a solution; it gives one a confidence that is lacking when one has only one solution. The two models may be equally good, but the two together are better than either one of them.

Why We Chose This Problem

We think this is a good introductory problem. What do you think?

Our experience is that an important stage in learning to model is the equivalent of jumping into the deep end of a swimming pool. Did you ever feel that you were floundering?

The swimming novice who jumps into the deep end knows what is required: to stay afloat and try to get to the side of the pool. Did this problem have clearly stated goals? We think so. The problem was stated simply and did not require much elaboration. The challenge was not in understanding the problem or in figuring out what kind of answer was required. The challenge was in getting to an answer!

Jumping into the deep end has an element of surprise for the uninitiated: There is nowhere to stand! This problem was designed to pose a similar dilemma. The goal of the problem should have been clear enough, but was there an obvious way of getting to that goal? We hope not. There is, for example, no simple formula that leads to an answer. We wanted to challenge you to *devise* a way of getting to an answer in the same way as the swimmer has to find a way of getting to the side of the pool. (In both cases it is a matter of finding the proper sequence of strokes.)

It does not help to encourage a novice to jump into the deep end and then promptly jump in yourself for the rescue. In the same way, it would not have been constructive if we had just shown you how to solve the problem. It is far better for us to do the equivalent of shouting out advice from the edge of the pool. The heuristics we have suggested and the questions we have asked are the modeling equivalent of "hold your breath," "shut your mouth," "can you float on your back?" and "have you thought of kicking your legs?"

One does not, of course, push people into the deep end for the sheer pleasure of watching them struggle. This is a much more purposeful initiation. The objective is to overcome the novice's fear. The swimmer does not really believe you if you say "Don't worry, you won't drown; you will manage to get to the side of the pool." The actual experience of doing so is the only way to build confidence: No pain, no gain!

The same is true of modeling. If you managed to get to the solution of this problem yourself, however much you may have struggled, we suspect that your confidence will have been bolstered.

This is a good problem for building confidence. The solution is not obvious, but there are a number of different lines of thought that all lead to an answer.

This is also a good problem because there is so much that can be learned in the process of getting to an answer. Did you notice how often we asked questions such as "What did you do?" or "Why did you do it?" or "Did you notice how this (or that) heuristic helped you?"

This was a good problem if it taught you to ask questions. Did it?

SIMILAR PROBLEMS

There are a number of problems that are similar to the gas tank problem. They are often referred to as "compartmental" problems. All involve the same type of thinking and can be solved in much the same way. Here are two examples:

Liquids and Containers

Figure 3.7*a* shows two vessels. They could have the same volume, or be of different volumes. The top vessel initially contains liquid A, while the bottom vessel initially contains liquid B. The two liquids do not react chemically. Suppose

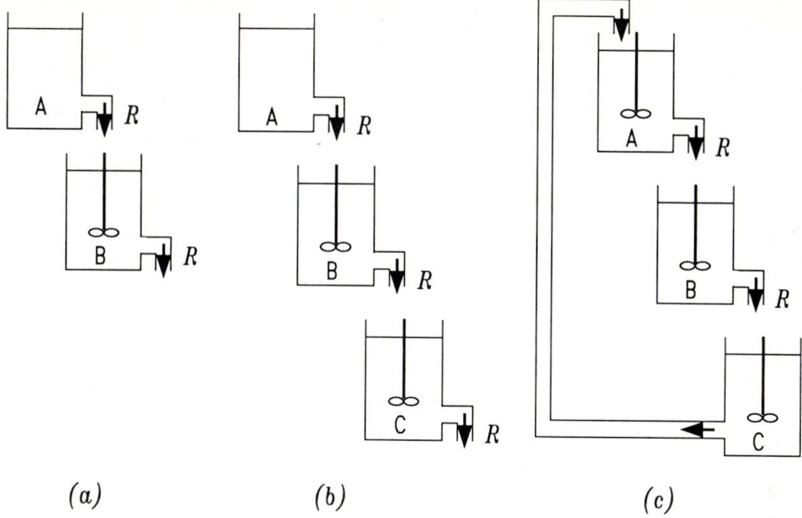

FIGURE 3.7
Three problems on liquids and containers: (*a*) two compartments; (*b*) three compartments; (*c*) three compartments with feedback.

that liquid from the top vessel flows into the bottom vessel at a constant rate R. The mixture in the bottom vessel is stirred continuously, and the mix flows out of it at the same rate R (so that the volume of liquid in the bottom vessel is constant). What proportions of the two liquids will be left in the bottom vessel when the top vessel is empty? Will this answer depend on the rate R?

If you want to stretch your imagination (and your spreadsheet), suppose that the mixture from the second vessel flows into a third vessel that initially contains liquid C. The liquid in the third vessel is also stirred and flows out at the same rate R, as in Figure 3.7*b*. What mix of liquids will be left in the third vessel when the first is empty?

Alternatively, suppose that the output from the third vessel is pumped back into the first, as in Figure 3.7*c*. How long will it take before there is more than 5% of liquid B in the top vessel?

A MEDICAL PROBLEM

Figure 3.8 represents an organ in your body, perhaps your heart (or brain or kidney). A dye or radioactive tracer is injected at a constant rate (for a limited time) into the artery that pumps blood *into* the organ. Small samples of blood are withdrawn at regular intervals from the artery that pumps blood *out* of the organ. These are stacked and labeled and subsequently analyzed to determine what concentration of dye or tracer they contain.

Predict the results of this experiment.

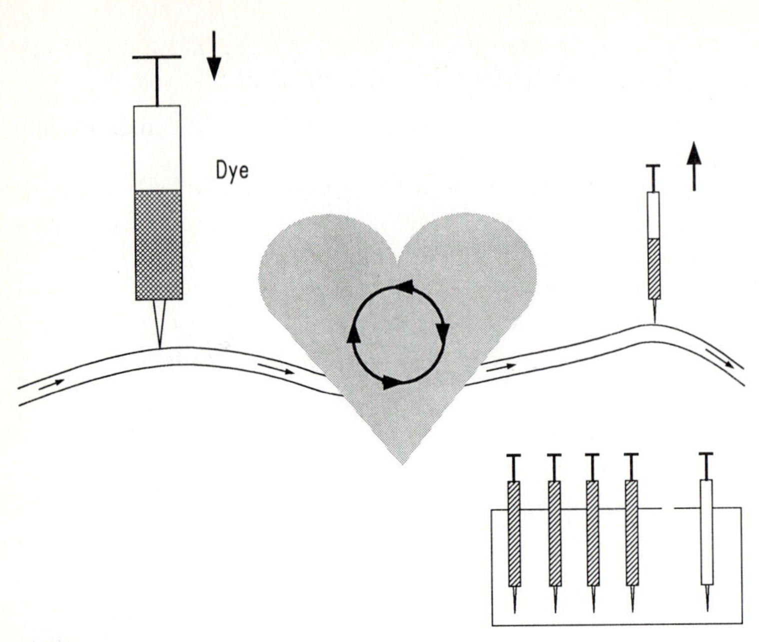

FIGURE 3.8
A medical test. Dye is injected on the left. Blood samples are extracted on the right at regular time intervals and stacked in sequence.

Alternatively, how could you use the results to determine the volume of blood stored in the organ?

FURTHER READING

The most complete discussion of heuristics that we know is *Discussion of the Method* by Billy V. Koen at the University of Texas at Austin. Koen describes engineering heuristics in five categories: rules of thumb and orders of magnitude, factors of safety, attitude determining heuristics, risk controlling heuristics, and resource allocation heuristics. Specific heuristics include

Make small changes in the state of the art
Allocate sufficient resources to the weak link
At some point in the project, freeze the design
Design for a specific time frame
Apply science where appropriate

Unfortunately this manuscript has not been published. The first three chapters are available in *Definition of the Engineering Method,* published by the American Society for Engineering Education in 1985.

Alan H. Schoenfeld's three questions to promote metacognition are contained in his excellent book *Mathematical Problem Solving* (New York, Academic Press, 1985). The book outlines his research and theorizing on the key aspects of mathematical understanding. He contends that these are

Resources and access to resources (the structure and organization of domain-specific knowledge)
Heuristic strategies (rules of thumb for making progress on difficult problems)
Executive control (aspects of metacognitive behavior: deciding what to pursue and for how long)
Belief systems (one's point of view regarding the domain)

THE CASE OF THE HOT AND THIRSTY EXECUTIVE

THE PROBLEM

You are a modeling consultant for a large corporation. One hot summer afternoon the president of the corporation decides to drive out of town to his lakeside cabin to recuperate from a day of hard decisions. Uppermost in his mind, as he approaches the cabin, is the image of a large glass of cold beer.

Alas, when he goes to the refrigerator, it is well stocked with orange juice but contains no beer. He frantically searches the kitchen and finds a dozen warm beers in a cupboard. He puts six of these into the fridge and then phones you. He briefly but expressively describes the situation to you and then says "I want to know how soon I can drink a beer! Drop everything you are doing and come back to me within the next 15 minutes with an answer." His tone of voice implies that your career with the company depends on a prompt and dependable response.

Here is your chance to make a good impression. Before he can hang up you ask, "Have you packed the beers in ice from the freezer?"

He is not impressed.

"The freezer isn't working. The fridge isn't working all that well either. I found a thermometer, and the temperature inside it is only 10°C. I don't need any smart-aleck suggestions. Just do what I say and tell me when I can drink that beer!"

He hangs up.

YOUR FIRST TASK

We are not going to ask you to solve this problem in less than 15 minutes, but we are going to ask you to work out how you *could* solve it in less than 15 minutes.

Your first task is to represent or describe the problem in simplified terms that clarify how you might tackle it. Make a list of assumptions and also of questions you would like to ask the president (you have his phone number). However, remember the president is irritable. It would not be wise to ask him too many questions, and it might be downright dangerous to ask him stupid questions. (Even sensible suggestions, like cooling the beers at the bottom of the lake, are unlikely to be welcome. We advise you to just accept the problem as the president has stated it.)

Some of the heuristics you have already met might help you in your search for an appropriate representation. In the next section we prompt your memory, but do not read on until you have gone as far as you can without our prompting. You should spend no more than 45 minutes on this task.

✔

HEURISTICS YOU MIGHT HAVE REMEMBERED

Did you try to rephrase the problem to make sure you understood it? In particular did you try to rephrase it in a way that helped you to think about the answer?

Did you try drawing simple diagrams of what was happening?

Have you pinpointed the crucial aspects of the problem? Have you thought of questions to ask that would confirm or modify your perception of what is crucial?

Did you imagine yourself inside the refrigerator and ask yourself what was going on around you? Did you perhaps imagine yourself growing cooler inside the can or bottle of beer?

Did you ask whether there were any physical laws to consider (such as conservation of matter or energy)?

Are there any simplifying assumptions you can make?

Is there an easier version of the problem that you might be able to solve?

Did you try to bound the answer?

Did you look for a ready-made formula for the answer?

If you thought of a model, did you ask yourself how it would help you? How would you actually use it?

Did you think of your solution as a number, a graph, or a table (or what)?

Did you try to identify essential variables? Did you choose a notation?

Did you use "salami tactics"? Did you try to develop a sequence of steps that would lead to the required answer?

Have you noticed that the above heuristics are a repetition, almost word for word, of the heuristics we discussed in the previous chapter? Have you found any of them useful for this problem? Has reiterating them helped you to think about this problem?

QUESTIONS, ASSUMPTIONS, AND THE GORDIAN KNOT

Do you have a list of questions for the president? For example, do you have questions about temperatures? Do you need to know how warm the beers in the cupboard are? Do you want the president to tell you exactly what he means by a cold beer?

If these are questions on your list, pause for a moment and anticipate the responses you are likely to get. Is there any use in asking the president these questions? Are you likely to get answers from him, or are you more likely to irritate him? Are there any assumptions you could make that are likely to be more useful than a phone call to him? Are there approaches you could take that bypassed asking him these questions?

Perhaps you want to know whether the beer is in bottles or cans. If it is in bottles, you might want to ask about the shape and size of the bottles, or the thickness of the glass.

Again, are these useful questions to ask? Is it essential to know the answers?

Have you been rummaging through your physics and chemistry books? Have you been trying to find the specific heat of beer or the thermal conductivity of glass? Or have you tried to find a book that tells you how a refrigerator works?

More importantly, have you asked yourself whether in 15 minutes you would have had the time to do any of these things?

The one book that you almost certainly did not consult was Plutarch's life of Alexander the Great. It would have been a good book to look at because it contains a relevant heuristic. According to legend, he who undid the Gordian knot would rule the world. Alexander went to see the knot, took one look at the tangled skeins, drew his sword, and cut through it.

A new heuristic is: If you are faced with a mess of details and you have to get to an acceptable answer in a relatively short time, then look for a way of cutting through the details.

Have you tried to cut through the details in this problem? Remembering yet again that you only have 15 minutes in which to find an answer, have you reduced the problem to its barest essentials?

OUR REPRESENTATION

Figure 4.1 is our representation of the problem. The warm beer is in a container (think of it as a sphere, if you like) which separates it from the cold interior of the fridge. Heat flows through the container from the beer to the

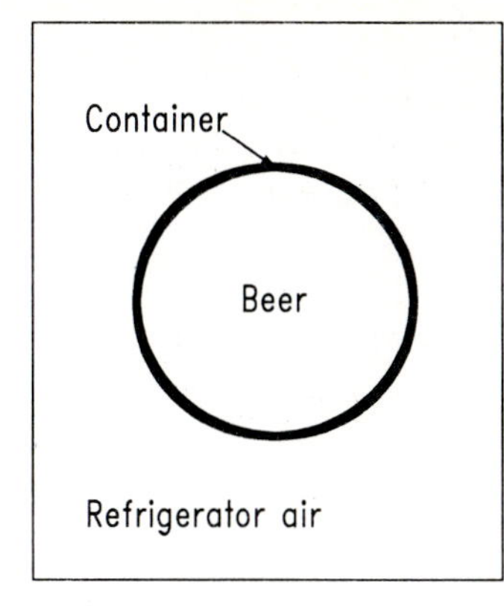

FIGURE 4.1
A line drawing showing the essential elements of the beer cooling problem.

fridge. As a result, the beer grows cooler, but the interior of the fridge does not grow warmer.

The essential elements of our representation are thus:

- A cold fridge that maintains a constant temperature
- A thermal barrier between beer and fridge
- Beer that is initially warm but passes heat through the barrier to the fridge

What assumptions have we made in our representation?

Our more important assumptions are that

- There is a device which removes heat from the interior of the fridge and a control which keeps it at a constant temperature. Even though the fridge is not working efficiently, it can remove heat as fast as the beer dissipates heat.
- The beer in the container is all at the same temperature.
- The shape of the container is unimportant.
- We can think of the container as an insulating barrier which affects the rate of heat flow from the beer to the fridge.

What are our variables? How do we want to represent our answer to the problem?

We prefer to think of our answer as a graph rather than a number. Our objective is to construct a graph, such as that sketched in Figure 4.2, showing how the temperature of the beer decreases with time. It follows that our important variables are time and the temperature of the beer.

Trying to sketch Figure 4.2 forces us to think about the shape of the graph. If we were sitting inside the beer, what would we "feel" that would make us grow cooler? What would "drive" the loss of heat?

The temperature of the beer will drop more slowly the closer it gets to the fridge temperature. When it actually reaches fridge temperature, it will not change

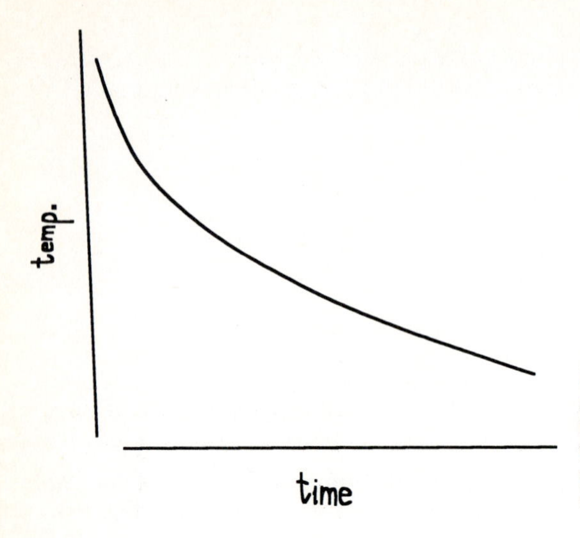

FIGURE 4.2
A rough sketch showing how the beer temperature might change with time. It is always useful to anticipate your results in a rough sketch or calculation.

at all. The difference in temperature between the beer and the fridge must therefore be an important "driving" factor. It follows that one would expect a graph of beer temperature versus time to be similar in shape to Figure 4.2.

What then is the role of the container?

Again, imagining ourselves inside the beer, we would grow cooler more slowly if the container were thicker or made of a more insulating material. The container must therefore modify the *rate* at which the temperature difference drives the heat loss.

If your representation was more complicated than this, could you have made it any simpler? How does one recognize when one has reduced a problem to its barest essentials? Have we cut through the Gordian knot? Have you?

YOUR NEXT TASK

You have an advantage over us, you know how we are thinking about this problem, but we do not know your representation. That means you are going to have to decide whether you prefer your representation as it stands or whether you want to modify it in light of ours.

Either way, your next task is to complete this assignment. That means you should

1. Decide on what information (if any) you are going to require from the president.

2. Develop your model in detail, assuming you have the information you need.

3. Implement the model, making estimates where required.

4. Decide on what you definitely need to ask the president and plan your conversation with him.

✔

OUR SOLUTION

Our Model

How to construct a graph such as Figure 4.2?

We know that at a time which we will call $t = 0$ the beer is at cupboard temperature. Let us call this C. We also know that after a very long time the beer will be at fridge temperature. Let us call this F (even though we have been told that F equals 10°C). Our problem is to track the drop in temperature from C to F over time.

We decided to follow the example of the previous chapter and try to compute the changes in beer temperature step by step.

Suggest a notation that will help us to do this.

Subscripted variables go hand in hand with step by step calculations. Suppose we choose a time step dt. Then we could let T_n denote the temperature of the beer after n time steps. It follows that we know T_0: $T_0 = C$, the cupboard temperature.

If we know T_0, can we find T_1? Can we perhaps write down a word equation that tells us how to find T_1 from T_0?

How about

$$T_1 = T_0 - \text{the drop in temperature during the time step } dt$$

What will that drop in temperature depend on?

We have already identified two factors:

1. The difference in temperature between the beer and fridge; in our notation that would be $(T_0 - F)$.

2. The insulating properties of the container.

If we pause to consider what else could be important, a further two factors emerge:

3. The time step dt: the longer the time interval, the larger the drop in temperature.

4. We have talked about a *drop in temperature,* but in physical terms we really should have been talking about a *loss of heat*. Heat loss is directly proportional to temperature drop, where the constant of proportionality depends on what physicists call the *thermal properties* of the beer. The fourth factor is therefore the thermal properties of the beer.

The first and third points suggest that we write

$$T_1 = T_0 - Kdt(T_0 - F) \tag{4.1}$$

where K is a constant of proportionality.

But how do we include the second and fourth points? We have already introduced an "extra" parameter K. Will K depend on the thermal properties of the beer? If so, will we have to introduce another parameter for the thermal properties of the container? The "barest essentials" implies as few parameters as possible. Could we perhaps "fudge" the thermal properties of the beer and container by thinking of them both as part of K?

It does in fact make sense to do so. We argued earlier that the container was a thermal barrier that altered the rate of heat flow from the beer to the fridge. This is precisely what K does. But K must also depend on the thermal properties of the beer. We will therefore "lump" all the physical constants we do not know into the single parameter K.

This is our equivalent of Alexander's sword!

Can we generalize Equation 4.1? Can we write down a formula that tells us how to step from temperature T_n (after n time intervals) to T_{n+1} (after n + 1 time intervals)?

Looking carefully at the logic behind Equation 4.1 suggests the generalization:

$$T_{n+1} = T_n - Kdt(T_n - F) \tag{4.2}$$

(Have you checked that Equation 4.2 reduces to 4.1 if you substitute $n = 0$?)

Equations 4.1 and 4.2 are easy to implement on a computer. If we do this sensibly, we will make provision for altering C, F, K, and dt whenever we wish to do so. In fact, once we have implemented the equations on the computer, our next problem is to choose values for C, K, and dt.

Estimating Numbers

Notice that we have not yet had to ask the president for any information. This is where we will probably need his help, but we will try to ask him for as little as possible.

We should be able to make a reasonable guess at C without risking a phone call to the president. We will suppose that, since it is a very hot day, $C = 35°C$. Even if we had not been told that the interior temperature of the fridge (F) was 10°C, we could have guessed that it was likely to be 5° or 10°C above freezing.

How to choose dt? What unit of time should we use for this problem?

It is reasonable to measure time in minutes for this problem. The smaller we make dt, the more accurate our model will be. (Why?) This suggests that dt should be a fraction of a minute. Since we have so little time to solve this problem, rather than argue about dt we will make it small, e.g., one-fifth of a minute.

That leaves us with the most difficult choice of all. How do we choose K? Do we have any way of calculating it? Do we have the time to calculate it? Or will we have to think of some way of measuring it?

Since the computer is ready to run as soon as we give it a value of K, it would be a good idea to see how K affects the answer. Figure 4.3 shows solutions for five different values of K. This figure is crucial. It highlights two important facts:

1. We can immediately discard values of K that are too high or too low. We know from experience that the first and fifth graphs are unrealistic. That is the good news.

FIGURE 4.3
A suite of cooling curves for different values of the parameter *K*. Intuitively, the top curve is too slow and the bottom curve too fast.

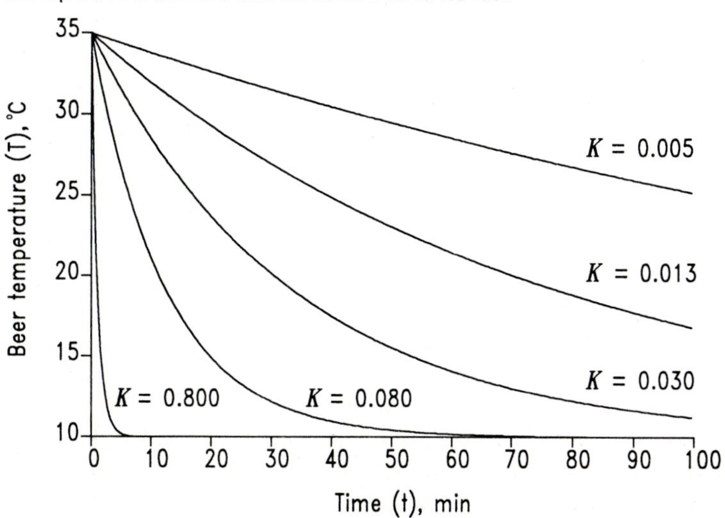

2. The bad news is that within the range of plausible values for K, the actual value of K makes a significant difference to our answer. We have to estimate K fairly accurately.

Time is running out. Here we are, sitting with this suite of graphs, wondering what the value of K should be, and it is almost time to make that phone call to the president. What do we do? What would you do?

Calibrating the Model

It should be apparent at this stage that we cannot guess at a value for K. Somehow we have to calibrate the model, i.e., look for evidence that suggests a value for K. This evidence will have to come from the "real world," and our only access to the real world is via the telephone and the company president.

While we have been developing our model, the beer has been cooling in the fridge. We know how long it has been in the fridge. If only we could estimate its *current* temperature, we could plot a point on Figure 4.3 (we would know both time and temperature). That point is likely to be close to one of our curves, which would help us estimate a value for K.

We know that the president has a thermometer. We could ask him to take a beer out of the fridge and measure its temperature. Even if he did not have the thermometer, we could have asked him to compare the beer in the fridge with the orange juice that has been there for a long time (we assume it is at fridge temperature) and the beer that is still in the cupboard. He should be able to tell us whether the fridge beer temperature is half way between the other two, or quarter way and closer to the cupboard temperature, or whatever.

However we do it, it is crucial that we find an estimate of at least one point on Figure 4.3. Once we have that point, we can quickly substitute three or four values of K in our computer model and look for two graphs that *bound* our solution, one that passes just above the point and one below it, as in Figure 4.4.

Presenting Our Solution

We still have not asked the president how cold he likes his beer. However, notice that Figure 4.4 allows us to have a far more intelligent and useful conversation with the president. For instance, we can make statements such as "You will have to wait another 30 minutes if you like your beer to be below 15°C, but can drink it in only 15 minutes if you will settle for 20°C." The president can react to this, ask us questions, and then decide for himself between various alternatives. (If the president did not have a thermometer we could still have offered him alternatives such as "If you wait another 15 minutes then the fridge beer will be half way between the orange juice and the cupboard

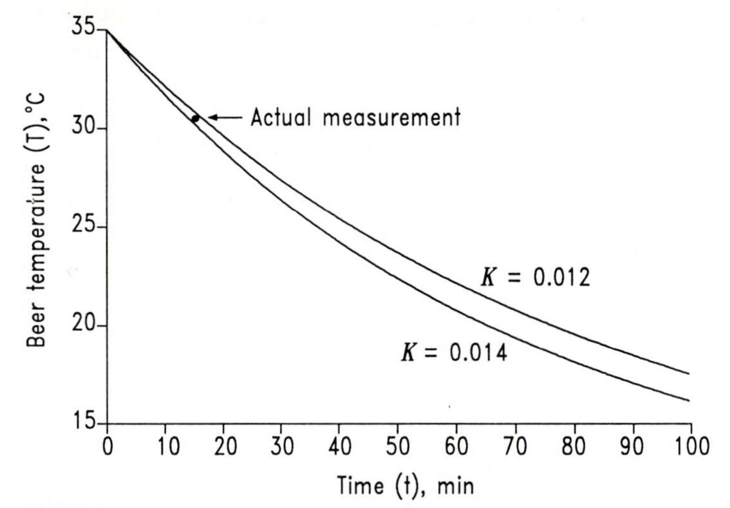

FIGURE 4.4
Actual measurement shows the beer temperature 15 minutes after putting the beer in the fridge. Once we know this, we can quickly generate cooling curves that pass just above and below the experimental point.

beer. On the other hand, you will have to wait another 30 minutes before beer temperature is one quarter away from juice temperature.'')

DISCUSSION

Heuristics and the Key Features of Our Solution

Identify the key steps in your solution to this problem. Then compare them with the key steps in our solution.

We think there were seven key steps in our solution of this problem:

1. Reducing the problem to its essential elements, as in Figure 4.1.
2. Deciding to present the solution as a graph.
3. Deciding to construct the solution one step at a time.
4. ''Lumping'' several factors into the single parameter K.
5. Generating a suite of graphs for different values of K.
6. Calibrating the model via information from the ''real world.''
7. Presenting the answer as a trade-off rather than a number.

Can you pinpoint the heuristics you invoked? Which of the heuristics presented earlier on were useful? Which do you think we used?

If you reread the heuristics we suggested, you should recognize that most of them were helpful. We drew a simple diagram; we imagined ourselves in the fridge or in the beer; we decided to present the solution as a graph, we identified time and the beer temperature as the essential variables and made assumptions about the temperature in the cupboard. We used "salami tactics," and we even bounded our solution right at the end of the exercise.

As in the previous chapter, a number of different heuristics would have pointed us in the right direction. There are, however, important differences between this chapter and the previous chapter.

For a start, we had a severe time constraint in this problem. This meant that we had to be exceptionally careful and ruthless in reducing the statement of the problem to its essentials. If we had gone wrong in the first step, if we had designed even a slightly more complicated representation of our problem, it is unlikely that we could ever have met our deadline.

This problem was also more complicated than the problem in the previous chapter. There were all sorts of considerations (such as the shape of the bottle and the thermal properties of the container and the beer) that were potentially important. However, we could not afford to dwell on their importance or even think carefully about the physics of the problem. This is where "cutting through the Gordian knot" was a crucial heuristic. It is one that you will find yourself using frequently, but you will have to learn to use it cautiously. To change the analogy, you have to be careful not to throw out the baby with the bath water.

The technique we used for cutting through this particular knot is also one you will use often. We "lumped" parameters. Instead of several parameters (such as the thermal properties of beer and glass or aluminum), we introduced one, the proportionality constant K. Can we lump parameters? is another heuristic to add to your list.

Why We Choose This Problem

This problem was a challenge: it was not trivial; it had to be solved in a short time; and the only link to the "real world" was an irritable company president.

Did you feel that this problem was unrealistic and/or a little too artificial? If so, why do you think we decided to include it in this book?

There are two ways of challenging the modeler. One is to present a complex problem. The other is to present a simpler problem but impose restraints on time and/or resources for its solution. We chose the latter approach because it makes it that much easier to present the problem. The tasks you were expected to perform should have been clear; the time (and information) constraints added the extra pressure.

Pressure serves a purpose. It rewards clear thinking above muddled thinking and good modeling above poor modeling. It creates an environment that highlights the heuristics one is trying to demonstrate. It creates a need for initiative and imaginative thinking.

Like the problem in the previous chapter, this is a good "deep end of the swimming pool" problem. You may struggle with the problem, but the struggle should be exhilarating and instructive.

While this problem may be somewhat unrealistic, it has the characteristics of a class of important real problems. Where else is one under pressure to get a solution quickly with only incomplete information? Weather forecasting is one answer. Can you think of others?

What We Like about Our Solution

How did you end up solving this problem? How does your solution differ from ours? Which do you prefer? Why?

Is your solution one that could have been developed and implemented within the 15-minute time constraint? Is ours?

We are not sure of the answer to that last question, but we are confident that our model is the meanest, leanest, simplest model that could have been built.

What do we mean by this?

We cannot subtract *anything* from our model without negating its whole purpose. There is nothing superfluous in it.

We have also been careful to ask the company president only for essential information. We could not have avoided asking the questions we did ask.

On the other hand, we managed to use our model skillfully. We like the fact that we did not present our answer as a single number (like 45 minutes). We also like the fact that we did not ask the president exactly what he meant by a "cold" beer. It was far more useful to present him with alternatives, to show him the trade-offs and then let him make his own decisions.

The model we built was simple, but the way we presented our answer made good use of the model and directly addressed the company president's problem. It is amazing how often modelers, who have spent months building and implementing a model, skimp on the conclusions they draw from the model.

Notice the important role that graphs played in both our approach to the solution and the way in which we were able to present the results. It nearly always pays to think of an answer as a graph rather than as a number. In fact in this problem we produced a series of graphs (for different values of K). That was also a sensible approach under the circumstances. It meant that we could react to the information we received from the president without running back to the computer.

What points would you like to have checked more carefully if there had been more time?

We would want to find out how sensitive our solution is to the choice of time step dt. Did we make dt sufficiently small? We would also want to know how sensitive our solution is to the temperature we assumed for cupboard beer. An even more important question is "If the president makes an error in estimating the temperature of the beer when we phone and ask him to do so, how does that affect the usefulness of the advice we give him?"

Under the circumstances there is little that we can do about these points, except to note that the upper and lower bounds give us some idea of sensitivity. This is one of the reasons why we chose to bound the solution rather than try to find a value of K that fitted the measured temperature exactly.

Incidentally, when we have presented this problem in class, some students have asked the president to estimate the temperature of the beer in the fridge after the first minute or even after the first time step dt. What is wrong with this? Why is it much better to ask him to estimate the temperature as late as possible?

An Alternative Algorithm

You might have used an approach to this problem that was similar to ours, but have developed a different algorithm. For example, you might have chosen "time" as your *dependent* variable instead of the temperature of the beer. Let us briefly develop this algorithm:

Imagine a series of steps. At each step we reduce the temperature of the beer by a fixed amount dT. After n such steps the beer temperature will be

$$T_n = C - ndT \qquad (4.3)$$

Let t_n be the time that it takes for the beer temperature to drop from C to T_n. Then obviously $t_0 = 0$, and by an inversion of the arguments we used to develop our algorithm, you should be able to argue that

$$t_{n+1} = t_n + \frac{kdT}{T_n - F} \qquad (4.4)$$

where k is a constant of proportionality that depends on the thermal properties of the beer and container.

Equations 4.3 and 4.4 could be implemented on a computer to produce graphs similar to Figure 4.3.

Our two approaches to this problem introduce new terminology. In our original approach we made calculations at regular time intervals dt. This is what is known as a *time-driven model*. In this alternative approach we made calculations only when the temperature had changed by a certain amount. This is what is called an *event-driven model*.

Which is the better model?

In the previous chapter we introduced two criteria for comparing models. The assumptions behind the two models in this case are identical so there is no way of discriminating between them on that count. The only question then is whether the one algorithm is technically better than the other.

Why should the event-driven model be better when the beer has just been put into the fridge (i.e., for small times)? Why is the time-driven model likely to be better for large times?

When Is a Chicken a Sphere?

An engineer, so the story goes, once hired a modeler to design a machine for plucking chickens. "How are you getting on with your design?" asked the engineer. "Very well indeed," replied the modeler, "I have assumed that a chicken is a sphere."

The above story has been told to one of us many times, and each time with the same mischievous relish, by an engineer who is healthily skeptical about modelers, computers, and engineers straight out of college.

However, it is a story that raises an important modeling question: When is a chicken a sphere?

The assumptions we make when building a model depend on the objectives of the model and the circumstances (e.g., available data) and constraints (e.g., time or money) under which we build it. An assumption that is good under one set of circumstances may be ludicrous under another.

This exercise was designed to illustrate these ideas. It is no coincidence that in our representation of this problem we say "The warm beer is in a container (think of it as a sphere, if you like)." A sphere has no special features; it is a neutral shape. We are implying that the shape of the container is unimportant.

Is this a good assumption? Can we change the problem in some way to illustrate when this would be a bad assumption?

It is fascinating to ask how our model might have looked if circumstances had been different.

Suppose, for instance, that a brewery is planning to boost its sales by designing beer bottles that really do reduce the time required to cool beer in a refrigerator. Under these circumstances we would have *days* rather than minutes to model the cooling process.

What difference would that make to our model?

Having changed the objectives, we have to go back and reconsider the important aspects of the problem. Some relevant questions are:

- Do cans cool faster than bottles? Always?
- How important is the temperature profile across the container wall?
- How will the shape of the container change the cooling process?

- Is the temperature distribution of the beer inside the container important?
- What about the surfaces of the container? What if the outside is wet?

The considerations that have the biggest influence would have to be identified using appropriate models. (Yes, more than one!) Some considerations may require only small changes to the model we built in this chapter. Others may require models that are orders of magnitude more complex. We might even end up using a supercomputer to study how drops of moisture form and run down the side of the container.

Why would one possibly build such vastly different models to describe essentially the same process for cooling beer in a container?

Because the *purposes* are different!

SIMILAR PROBLEMS

The beer bottle problem is an example of a "rate" model. We do not know how the temperature of the beer changes directly, but we can speculate about its rate of change: how much the temperature will drop in a short time interval. Rate equations are important in physics, biology, economics, medicine, and even (as the following example shows) in politics.

A Rumor in a Small Town

Y and Z are both candidates for mayor of a small town (population 5000). Five conspirators in Y's campaign headquarters agree to besmirch Z's reputation. Whenever the opportunity arises in conversation with their fellow townspeople, the conspirators will casually mention that Z has a serious drinking problem. Of course, everybody who hears this believes it. What is more, everybody who believes it helps to spread the rumor.

You are a political analyst. You notice what is happening. You also know that at the start of this conspiracy, 4 out of every 5 voters were planning to vote for Z. The election is in 4 days' time. Analyze whether or not Z is still likely to win.

Alternatively, you work at Z's campaign headquarters. You only discover what is happening 2 days before the election. You call an urgent meeting of your campaign staff (all six of you) and start refuting the rumor. Each time you meet somebody, you will say "Has anybody told you that Z has a drinking problem? You should know that Y has been spreading this rumor and that it is totally unfounded." Well, you can take the problem from there!

An Epidemic

It does not take much imagination to turn the previous problem into an epidemic problem. For example, two students in a school of 2000 come back from their vacation incubating measles. If the authorities recognize that there is an epidemic only after so many weeks, and can only inoculate so many students a day, how many students in toto will contract the disease? How does your answer depend on the number of students who had been previously inoculated or who were already, for whatever reason, immune?

Contamination from Radioactive Waste

In the year 1980 a container of radioactive substance P was buried in a mine. Now, P decays into another (and far more dangerous) radioactive substance Q, while Q decays into a stable (and benign) substance R. In the year 1990, a government agency takes a sample from the container and finds that its composition is 90 percent P, 8 percent Q, and 2 percent R. You have been hired to predict the percentage of Q in the container over the next 1000 years. In particular, when will there be a maximum amount of Q in the container?

FURTHER READING

The beer bottle problem is also an exercise in making assumptions. In *Consider a Spherical Cow: A Course in Environmental Problem Solving* (Los Altos, CA, Kaufmann, 1985), J. Harte focuses on approximation and modeling to cope with word problems that require a quantitative answer. The book avows to help readers develop a knack for stripping away unnecessary detail. Harte asserts that the text "should teach the reader how to transform realistic, qualitatively described problems into quantifiably solvable form and to arrive at an approximate solution."

Thinking Physics: Practical Lessons in Critical Thinking by L. C. Epstein (San Francisco, CA, Insight Press, 1987) is a delightful book in which questions appear first, followed by explanations—just as problems appear in real life. The reader is encouraged to read a question and to stop and think. Epstein claims "the most important problem in physics is *perception,* how to separate the nonessentials from the essentials and get to the heart of a problem, *how to ask yourself questions.*" Despite the snide comments about physics in this chapter, it would appear that physics and modeling have much in common.

CHAPTER **5**

TENNIS, ANYONE?

A SURE BET?

Pat and John have dominated the world tennis scene for a number of years. They are to be found at every major tennis tournament, locked in battle and competing for titles. As an avid tennis fan, you and your friends have followed almost every match between them.

Your best friend is an admirer of Pat; you think John has more style. However, even your friend concedes that John has an impressive service. You have been collecting all sorts of statistics on John and so you *know* just how impressive it is. Your records show that John has served to Pat 230 times and has won 154 of those points!

Pat and John will be facing each other at Wimbledon next week. Your friend suggests that since you are so sure of John's prowess, you should risk a wager. You craftily suggest that you will bet only on those games where John has the service. "Sure," says your friend, "Each time John wins a game, I will give you $3, but each time Pat wins you will have to give me $10!" (It looks as though your friend has been collecting statistics too.)

Is this a bet you should take?

To answer this question it is important for you to know something about the scoring system in tennis and to understand the difference between a "point" and a "game." We explain this in the next section. If you are already familiar with the basic rules of tennis, skip the next section and move on to "Your First Task."

KEEPING SCORE

A tennis match consists of a number of "sets," and each set consists of at least six "games." Since you are betting only on the outcome of certain games, we will not discuss "sets" and "matches" here.

The same player serves the ball throughout a game. Your bet relates only to those games where John will be serving.

A game consists of a number of points. Each time John serves, a *point* is won or lost. The score begins at "love all." If John wins the first point, the score will be "fifteen–love." (Since John is serving, his score is given first.) If Pat wins the first point, the score will be "love–fifteen."

The scoring steps are: fifteen, thirty, forty, *game*. If John, for example, wins the first two points, loses the third, and wins the fourth and fifth, the score will be:

fifteen–love
thirty–love
thirty–fifteen
forty–fifteen
game: John has won that game!

The only complication occurs if the score reaches forty–forty. The next player to win a point does not win the game. Instead, at that stage a player has to win two consecutive points to win the game, which goes on (sometimes on and on) until that happens. The score forty–forty is called *deuce*. Whoever wins the next point is said to have the *advantage,* or "ad." If a player who has the advantage wins again, that player wins the game. Otherwise the score goes back to deuce.

For example, consider a game with the following scores:

fifteen–love
fifteen all
fifteen–thirty
thirty all
thirty–forty
deuce
advantage, John
deuce
advantage, Pat
game to Pat: Pat wins!

Confirm that John wins the first point, Pat wins the next two points, John the next, then Pat, two points to John and finally Pat wins the last three points.

YOUR FIRST TASK

Now let us go back to the question: Should you accept the bet?

Your first task is to give the best answer you can in not more than 10 minutes.

What have you decided? Are you willing to take the bet?

Did you really manage to analyze the proposition in less than 10 minutes? Or is your answer intuitive?

If you decided to accept the bet, why do you think it is a good bet? If not, why not? Write down or explain the argument behind your decision.

YOUR NEXT TASK

Your next task is to develop a more convincing argument or approach, one that takes into account the following questions:

1. On what basis will you accept or reject the terms of the bet? What is the rationale behind your decision?

2. What do you really need to know about the game of tennis? How can you use this information?

3. What data do you have about the performance of the two players? How can you use those data?

4. If you were given time to build a model, what type of model might be helpful? What could you expect to gain from it? What sort of answer or answers would you be looking for?

You should take about 20 to 30 minutes to consider these questions.

Incidentally, would you have asked these questions of yourself if we had not asked them? Why do you think they are good questions to ask at this stage?

✔

POINTS YOU MIGHT HAVE RAISED

What was your rationale for accepting or rejecting the bet?

It is not easy for us to anticipate your answer to this question. We are fairly sure that you would not accept if you were likely to lose money, but would you accept if you were likely to make money? That depends on your relationship with your best friend. We will assume that you do not mind taking a few dollars off a friend who is in fact trying to outsmart you—in an amicable sort of way.

It appears the odds are loaded against you: you pay $10 every time John loses, but get only $3 if he wins. What percentage of games would John have to win to make this a paying proposition?

Suppose John wins x games out of 100. Then your profit would be what you win on x games minus what you lose on $(100 - x)$ games, i.e.,

$$\text{Profit in dollars} = 3x - 10(100 - x) = 13x - 1000$$

We have decided that you want your profit to be positive, i.e., you want

$$13x - 1000 > 0$$

or

$$x > 1000/13$$

This tells you that John needs to win more than 77 percent of the games you bet on if you are to make a profit.

Did you make this calculation during the 10-minute exercise? If not, it is a calculation you should have made.

Now that you know how many games John would have to win to make a profit, the next question we need to ask is "How many games is John *likely* to win?"

This leads naturally to the question "What do we know about the performance of the players?" This is where your habit of keeping records pays off. You know that John has won 154 points out of the 230 times he has served to Pat. (Remember, the bet applies only to those games where John is serving.) And therein lies the crux of this problem. You have data about how many *points* John is likely to score, but you are betting on how many *games* John will win. You need to convert your information about points into information about games.

Did you realize this during the 10-minute exercise? Did you have time to think about how you might be able to make the conversion from points to games? If so, congratulations!

But what did you do under pressure of the time constraint? Did you just ignore the difference between points and games? Did you argue that since John wins only 154 out of 230 points (i.e., 67 percent of points), he is unlikely to win 77 percent of games? Did you decide not to take the bet on this basis?

Whichever way you argued, what assumptions did you make? Did you assume that a player is likely to win fewer games than points? Or the same? Or more? Did you have any arguments to support your assumption? If not, did you make a mental note to check the assumption at some stage?

It was unfair of us to expect you to make a reasoned decision in less than 10 minutes. Our objectives in setting the exercise were

1. To see whether you would calculate the percentage of games you needed to win in order to make the bet attractive.

2. To see what assumptions you would make under pressure.

3. To point out that your assumptions were probably unfounded. There really is no way to make a useful assumption in 10 minutes. This is a problem that has no quick and easy answer.

A good answer (unless you are a computational genius) to the 10-minute exercise is that you need more time to analyze the problem.

What would you do if you had more time?

We would try to make more careful calculations. Perhaps one should build a model.

And the purpose of the model?

To compute the percentage of games a player is likely to win, given information about the percentage of points that same player is likely to win.

What kind of model would that be? What would it depend on?

To answer the second question first: it would have to depend on the scoring system in tennis. To answer the first question, we need to look a little more closely at just what we mean by the word "likely" which we have been using in important questions such as "How many games is John likely to win?" and statements such as "We will accept the bet if we are likely to make money on it."

Where does chance play a role in this problem? You expect a bet to be a gamble. Where is the element of risk?

✓

History tells us that John has won 154 out of 230 service points. This means he has won nearly 2 out of every 3 points in the past, but what does that tell us about the *next* point? Will he win or lose it? If you are not already familiar with the concept of *probability,* now might be a good time to introduce you to it.

AN INTRODUCTION TO PROBABILITY

Suppose that you have 50 pieces of paper in an urn and that only 2 of those pieces are blank. The probability of putting your hand into the urn once and drawing out a blank piece of paper is 2 in 50, or .04. A *probability* is a number between zero (impossible) and one (certain).

Notice that you could determine this probability experimentally. Each time you drew a piece of paper it would either be blank or not, but if you put the paper back, shook the urn, and then drew again, you would expect in the long run to draw a blank very nearly 4 percent of the time.

What is the probability that John will win a point?

It looks as though the answer should be .67, or approximately 2 out of 3, or 2/3, but let us make sure that we understand where this comes from. You have kept score over a long period of time. During that time, John has won 154 out of the 230 points where he has served to Pat. On average, therefore, he has won approximately 2 out of 3 service points. If you look at your data more carefully, you will find games where John's performance was poorer than this, others where it was better. There might even have been games where John won every point. However, the best we can do with your data is conclude that his probability of winning any point is 2/3.

But that does not tell us whether he will win the next point. It only tells us how likely he is to win it. When we use the word "likely," we really mean that we need to estimate a probability.

Why do you think we talked about "a long period of time" in the above paragraphs? How confident would you have been if you had observed John for only one or two games?

Intuitively, any estimate of probability must improve as we make more observations. If we had based our estimate on only two games, for instance, we might have chosen two games where John just happened to be playing exceptionally well (or poorly) or where he was unusually lucky (or unlucky). It is important that we have data from many games.

If you do not trust your intuition, do an experiment. Flip a coin 32 times and count the number of times it lands heads up. Plot the ratio of heads to total flips after 2, 4, 8, 16, and 32 flips, as in Figure 5.1. Repeat the experiment. Repeat it again and yet again.

What do you observe?

Unless you have been cheating, or your coin is biased, you should notice three things. First, that the answer bounces around from one experiment to another: after two flips the ratio is 1 in Figure 5.1a, .5 in Figure 5.1b, 0 in Figure 5.1c, and .5 in Figure 5.1d. Second, the bouncing around decreases as the number of flips increases: after 32 flips the range across the four graphs in Figure 5.1 has narrowed to between .47 and .53. Third, the answer is getting closer and closer to .5.

The "bouncing around" from one experiment to another is related to what statisticians call the variance. Figure 5.1e illustrates the statistical result that the variance decreases as the number of experiments or instances increases. The more instances we have, the more confident we can be of our estimate.

All this tells us something about the algorithm we need for deciding whether or not John wins a point in the model we are going to build. A good algorithm must be one that predicts *on average* that he will win 2 out of 3 points. If, for example, we modeled 600 of John's services (notice we have chosen a large

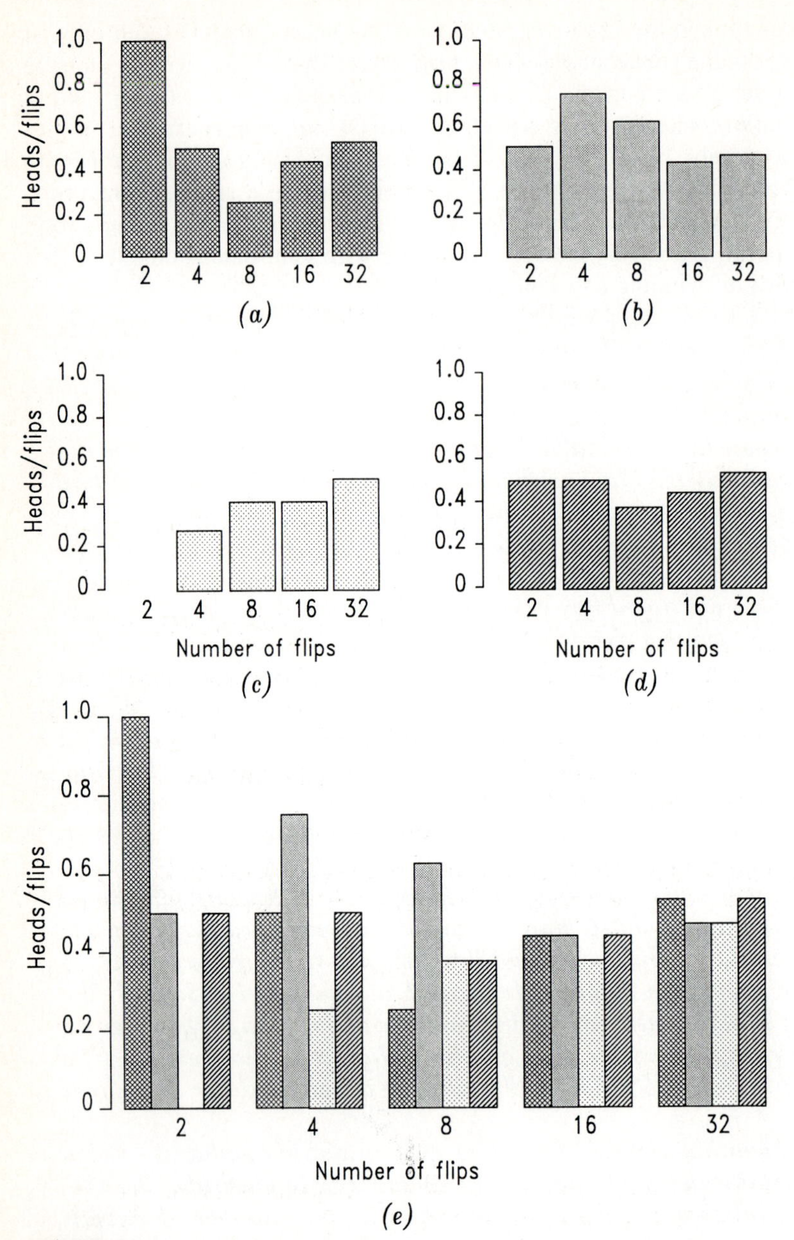

FIGURE 5.1
The ratio of heads to total number of flips gives an estimate of the probability of flipping heads. Plots (*a*) through (*d*) show four series of up to 32 flips each. When we put them all together, as in (*e*), it is easy to see how unreliable the estimate of the probability is after only 2 flips. We can be a lot more confident about the estimate after 32 flips.

number!) and our algorithm told us that he won 400 (or very close to 400) of them, then we *might* have confidence in the algorithm.

Suppose we have an algorithm with a repeated pattern of wins and losses, e.g., win, win, lose, win, win, lose,... This would indeed predict that John wins 400 out of 600 services. Should we have confidence in it? Is it a good algorithm?

Surely not! The very fact that there is a pattern introduces a bias. In fact applying it to a game would lead to the following scores:

John wins a point	fifteen–love
John wins a point	thirty–love
Pat wins a point	thirty–fifteen
John wins a point	forty–fifteen
John wins final point	*game*!

Pat will win the first point in the next game, but John will still win the game. In fact John would win every game. See what we mean by bias?!

(Can you think of a pattern that would let Pat win sometimes?)

We therefore have another essential requirement for a good algorithm. It must have no built-in pattern. In other words, it must be *random*. We should not be able to predict beforehand when the algorithm will let John win a point. But, going back to our previous requirement, the algorithm must still ensure that John wins approximately 400 out of 600 services.

How can we develop an algorithm with both requirements? Remember it must be random (there is no pattern to tell us ahead of time whether the algorithm will say "win" or "lose") but it must still ensure that John wins approximately two out of three points in the long run.

As a clue, suppose you had an unbiased die. How could you use it to decide whether John wins or loses a service?

✓

You could, for example, roll the die and say, "John wins unless the die shows five or six, in which case Pat wins."

Is this the kind of algorithm we are looking for?

Yes indeed. It is random, but it behaves in the way we want it to behave when we apply it many times.

We can achieve the same effect in a computer language or on a spreadsheet by using a *random number generator*. This is a built-in function or algorithm that will, for example, generate any real number between zero and one with equal probability.

Suppose that the random number generator produces a number z. Would you be happy with a rule that says: If $z < .67$, then John wins, otherwise Pat wins?

Incidentally, how would you test to see whether the random number generator we have been talking about is doing its job properly?

DESIGNING A MODEL

With this background you should be able to go ahead and design a model. Remember, the purpose of the model is to predict the percentage of games John is likely to win. (In other words, it will calculate the probability that John will win a game.)

While you are designing your model think carefully about how you will actually use it.

✔

OUR MODEL

We are trying to make a connection between a tennis point and a tennis game. If we think about it, a game is just a series of points plus a scoring system. In the section on probability we discussed how we could model a point using a random number generator. To model a game we just need to add the scoring system!

This suggests the following algorithm:

1. Initialize the score at "love all."

2. John serves. Use the random number generator to decide who wins the point.

3. Update the score.

4. Check the score to see if the game is finished. If it is, record who has won.

If not, go back to step 2.

This model in a sense is rather like "playing tennis on the computer." It follows the scoring rules of tennis exactly. The only difference between our model and a tennis game is that we have used the random number generator to decide who wins each point instead of playing the game stroke by stroke.

How does your model compare with ours? Did you also try to keep score on the computer?

There are two questions that we still need to answer:

1. How do we implement this model on the computer?

2. How do we use the results?

We will leave you to sort out the first question (if you have not already done something similar on the computer). The main complication, of course, is the way the scoring changes after deuce. But why solve the problem on a computer? Why not use a die and solve it with pencil and paper?

The second question is more fundamental. Suppose we implement our model on a computer and run it, and then it tells us that Pat has won the game; what do we do then?

How are you going to use your model? What should we do with ours?

✓

DRAWING CONCLUSIONS FROM STOCHASTIC MODELS

There is an important difference between this model and previous models in this book.

Previous models were *deterministic*. By this we mean that there was no element of chance built into the models. Unless you changed one or more of the input parameters, the model would always give you (or anybody who used it) the same answers.

However, if you run this model again, you could get a completely different output. The first time it might tell you that Pat won after a struggle, with the advantage first going to John, then to Pat, and then to Pat again. The second time it might tell you that John won resoundingly; Pat was unable to score a single point.

We call this type of model *stochastic*. Chance is built into it. Each time we run the model it is as unpredictable as a single spin of a roulette wheel or throw of a die. Only the average behavior of many runs of the model is consistent and predictable.

Since chance is an important part of our original problem, this was the right kind of model to build. We would have been worried if our model predicted exactly the same scores every time we ran it. (There would have been nothing to bet about!) But we are still left with the difficulty of drawing conclusions from our model.

What is the secret? How do we interpret the results?

The secret is to run the model many times and accumulate results. (This is one of the reasons why you needed to implement your model on a computer instead of using a die. The other reason was that the die would be difficult to use if we changed the input data; if John won 3 out of 5 games, for example.)

If we make repeated runs of our model, we will be able to make statements such as "John won p games out of q." If q is large, we will be able to conclude that "The probability that John will win a game is close to p/q."

We use the word "replicates" to describe repeated runs of a model where the only difference between one run and the next is what the random number generator produces.

How often do we need to run the model? Will 100 replicates be good enough? Or will we need 1000 replicates?

A statistician could advise us, but suppose we do not have access to such expertise? We could investigate for ourselves. For example, we could do a computer experiment that is similar to the coin flipping experiment illustrated in Figure 5.1. In this case we could run our model 1000 times, for instance, and record our estimate of the probability of winning a game after 100 games, 200 games, and so on. Then we could do it all again, and again. On the basis of that kind of experiment, we decided to run our model 500 times—because the variance was sufficiently small at that stage.

What is the probability that John will win a game? If you have not already done so, run your model until you are confident in your answer.

✔

WILL YOU TAKE THE BET?

We ran our model 500 times, i.e., we *simulated* (another modeling word meaning we used a model that was built to follow, step by step, the essential occurrences in the real world) 500 games in which John always served to Pat. John won 426 times, i.e., about 85 percent of the games. We have already calculated that you should break even if he wins 77 percent of the games. It looks as though you should take the bet!

Or should you?

The above calculation suggests that you will make a profit in the long run, but how many games will John and Pat actually play during a Wimbledon match? And in how many of those will it be John's turn to serve?

Remember, in the section on how to score in tennis, we said that since we were only betting on games, we did not need to know about sets and matches? To answer the above questions, we *do* need to know more about the rules of the game.

It is not unusual for information that seems irrelevant at the beginning of a modeling exercise to turn out to be important at a later stage, as our understanding of the problem improves.

The men's tennis match is the best of five sets. Depending on who wins the first three sets, a match might therefore consist of three, four, or five sets. A set is won by the first player to win six games, but there is a catch: The winner must lead his opponent by two games. Six games to four wins the set, but six games to five means that the players have to play at least one extra game. A set cannot go on indefinitely, however, and there are rules about "tiebreakers" that we will not go into here.

It follows from the above that John will play at the very least 18 games, and realistically more like 30 games against Pat during a typical match. He will serve in half of them, so you will be betting on the outcome of about 15 games.

How much money do you expect to take off your friend?

Eighty-five percent of 15 games is 12.75 games. So you should win 12 or 13 out of the 15 games.

Notice that it makes quite a difference whether you win 12 or 13 games.

What happens if you win only 11 games?

These calculations should suggest new and more sophisticated questions for you to ask.

What questions would you want to ask next? How would you use your model (or perhaps modify your model) to answer them?

EXERCISING THE MODEL

We would want to ask two additional questions before we take the bet:

1. How *sensitive* is our result to John's performance. What happens if John has an off day and his service is not quite as good as usual? What happens if it is a little better than usual?

2. We have used the model to tell us how many games John is likely to win in the long run, but we have realized that we will only bet on about 15 games during the Wimbledon match. If John wins 12 or more out of 15 games we will make a profit; otherwise we could lose money. Can we calculate the *risk* we are taking? What is the probability that we could actually lose money on those 15 games?

Sensitivity Analysis

Looking more closely at your records, you notice that on a poor day, John's success rate when he is serving drops by about 10 percent and that on a good day it goes up about 10 percent. In other words, his probability of winning a point on a poor day goes down from .67 to .60, while on a good day it goes up to .73.

To find out how many games John might win on a poor or good day, all we need to do is change the decision point on our random number generator from .67 to .60 or .73. However, notice one of the disadvantages of stochastic models: each time we change a parameter value we have to repeat a whole set of replicates. If we decided 500 replicates were sufficient originally, we have to run our model another 500 times with the reduced probability of success and yet another 500 times with the increased probability of success! It sounds like a lot of work, but only takes a minute or two on the computer, and the results we obtain are interesting. On a poor day John is only likely to win 73 percent of games, while on a good day his probability of winning a game goes up to .93!

Does it still look like a good bet to you?

Small Samples

Figure 5.1 showed us, in a different context, that the outcome of a stochastic process can have a high variance (i.e., bounce around a lot) if we take too small a sample. If we were able to bet on the outcome of hundreds of games between Pat and John, then we could feel comfortable making a decision (to bet or not to bet) based on the results we have obtained from our model so far. The model suggests that we take the bet.

But we are only going to bet on about 15 games. That makes it a more risky proposition; our winnings could be higher than expected (John might win 14 or even 15 games) or we might actually lose money (John might win 11 games or less). It would be useful to be able to quantify that risk. (We like to take a bet now and then, but we don't care to lose money recklessly.)

How could you use your model (or our model) to calculate the probability of losing money on a 15-game bet?

✓

We would simulate a large number (say 1000) of Wimbledon matches. Each match would consist of the 15 games where John has the service. We would calculate the percentage of those matches where, on balance, we lost rather than made money. Our algorithm would be as follows:

1. Let T be the running total of matches where we came out losing money. Set T equal to zero.

2. Run the model 15 times to represent 15 games. Calculate the amount of money made or lost on each game.

3. Add up all winnings and subtract all losses. If the answer is negative (a net loss), add one to the running total T.

4. Repeat steps 2 and 3, say 1000 times.

5. The probability of losing money when you bet on a sequence of only 15 games will then be approximately $T/1000$.

Look at step 2 carefully. Do you really need to run your model 15 times?

Not if you have already used it to calculate the probability that John will win a game. It is much easier (and quicker) to use a random number generator instead of the model. For example, if we know that John will win 85 percent of all games, we can simply say that he wins if the random number is less than or equal to .85; otherwise he loses. All we need do is generate 15 random numbers.

But what if John is having a bad day? Or a good day?

We decided to repeat the whole exercise three times: for his normal performance, a poor day (73 percent chance of winning) and a good day (93 percent chance of winning). Table 5.1 shows the results we obtained.

Notice how much information we get out of Table 5.1. The probability of losing money on a bet over 15 games is a much better guide for decision making than the average number of games John will win. Notice too how useful it was to repeat the exercise for good and poor days. This sensitivity analysis provides us with a broad perspective of the risks involved in the bet.

How you use the information in Table 5.1 to reach a decision is up to you. One of the attractions of a stochastic model is that it can generate all sorts of interesting information about risk. For example, think about how much more information you would want to have if this were a decision involving millions of dollars. How could you use your model or our model to generate that information?

TABLE 5.1
THE PROBABILITY OF LOSING MONEY IN A 30-GAME TOURNAMENT

Probability that John wins point	Probability that John wins game	Probability that you lose money
.60	.73	.61
.67	.85	.18
.73	.93	.02

*Based on betting whenever John serves (15 games).

Did you remember to ask how sensitive your answer would be to the number of games played during the match? Fifteen was only a rough approximation. Perhaps you need to find out the rules for tiebreakers after all!

DISCUSSION

The purpose of this exercise was to introduce you to stochastic modeling.

Notice that we did not set out by telling you, the reader, that this was going to be a stochastic modeling exercise. We set out with a problem and asked you to think about how to solve it. In the course of helping you to think (or echoing your thoughts, perhaps), we nudged you toward the concepts of probability and stochastic modeling. It was inevitable that you would use these concepts; there is no other way to analyze this problem!

Probability is not an easy concept to grasp. The idea that you can be more confident of the outcome of 100 games than 15 games, for example, may not be intuitively obvious. The exercise has been designed to give you the opportunity to appreciate these ideas through your own efforts.

Now that you have worked with a stochastic model, there are a number of questions that you probably want to ask about this type of modeling. We will try to anticipate and answer some of them here.

When to Use a Stochastic Rather Than a Deterministic Model?

We hope we have convinced you that there is no way you could have tackled this problem without bringing in the element of chance. In this case, therefore, the choice was obvious: You had to build a stochastic model.

Other cases are less clear-cut. Why not build a stochastic model of the gas tank problem in Chapter 3, for example? We could imagine a stochastic model. We might, for instance, build a model that feeds nitrogen into the tank in very small packages at regular time intervals. The model could keep track of each package. We could develop arguments that tell us the probability that a package will stay in the tank (or be expelled from it) during each time step. A random number generator would determine what actually happens to each package at each time step.

Should we have built a model like this?

We could have, but there would have been nothing to gain from it. Chance is not an important part of the gas tank problem. It is therefore a waste of time to build a stochastic model. Remember, we would have had to run the model many times to determine its average behavior. The stochastic details only confuse our understanding of the problem.

The rule is: Build stochastic models only where chance is an important part of the problem.

Even with this rule, there are still cases where it is not obvious whether to build a stochastic or deterministic model. The problem of Pat and John at Wimbledon illustrates this.

Suppose we had already calculated that John was likely to win 85 percent of all the games in which he served. Then there is a very simple deterministic model that we can build to calculate the profit that we *expect* to make if we bet on N such games. It is:

$$\text{Expected profit in dollars} = .85N \times 3 + (1.00 - .85)N \times 10$$

Notice that this model *only* tells you what you expect to win. Because it is deterministic, it will always give the same answer. It cannot give you information at all about your chances of winning more than that amount or even about your chances of losing money. In fact the deterministic model does not even *see* the possibility that you might lose money.

Now, if you are betting on not just one match at Wimbledon, but every game on the tennis circuit this year, then N is likely to be a large number (say 80 games), and this deterministic model would be adequate. For a large number of games the variance is small and can be ignored. In other words, chance events will balance one another.

But, as we have seen, if N is only 15 games or even less, the variance cannot be ignored. The deterministic model then only tells a part of the story. You need a stochastic model to give you the whole story. Only a stochastic model can provide additional information (very important information) such as the risk you run of losing money on the match.

How Detailed Should a Stochastic Model Be?

As with any model, the appropriate amount of detail depends on the purpose of the model and constraints of time, data, and resources.

The above discussion, where we started with data about games rather than points, illustrates this. We still wanted to build a stochastic model, but it is a simpler, less detailed model than the one we originally built. Points do not come into it.

Alternatively, if you had the data and the incentive you could build a more complex rather than a simpler stochastic model of John playing Pat.

You might, for example, know the probability that John will serve an "ace" (i.e., Pat will be unable to return the service so John wins the point) and the probability that he will misserve (so Pat wins without having to hit the ball). If the game is not decided on the service alone, you may have data that tells you the probability that John will return a shot to Pat or miss the shot, and similarly for Pat.

With these data you could build a model that decides each point, ball by ball. You really would be playing tennis on the computer. It would be fun to build such a model, but you still have to ask whether you are justified in constructing so detailed a model.

Looking for a More Direct Solution

The major part of our model is predicting how many games John will win, given that he has a 2/3 chance of winning each point. We tackled this by building a simulation model and running many replicates. There is an alternative approach that you might have explored.

Consider Figure 5.2. It shows (from John's point of view) possible outcomes during a game. For example, starting at the score of ''love all,'' there are two possible outcomes: fifteen–love and love–fifteen.

But one outcome is more likely than the other. What is the probability of a fifteen–love score? And a love–fifteen score?

The answers are 2/3 and 1/3, respectively.

What about a fifteen all score?

This is a more complex calculation. There are two ways to reach that score. Either the score is fifteen–love and John loses the next point, or it is love–

FIGURE 5.2
Showing, from John's viewpoint, all possible scores and outcomes during a tennis game. The score 30–30, for example, can be reached either from the score 30–15 or from 15–30. What might happen if the score reaches deuce (40–40) is shown in the next figure.

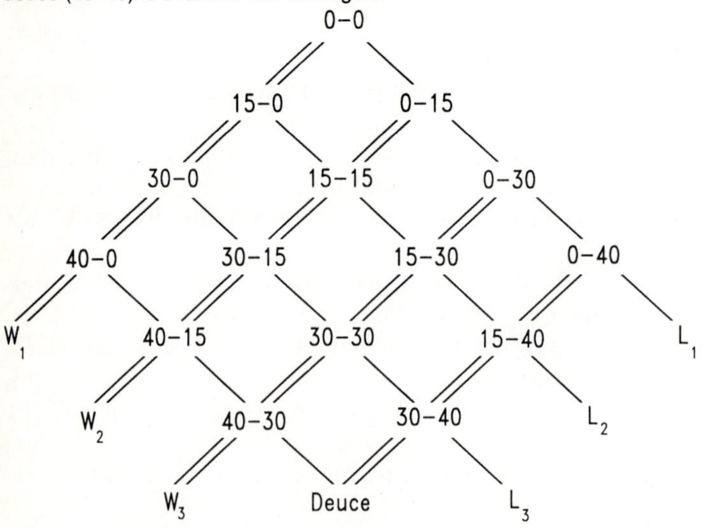

fifteen and John wins. The probability of a fifteen–love score, as we have seen, is 2/3, and the probability that John loses the next point is 1/3. The probability of both events (fifteen–love and John loses the next point) is therefore 2/3 times 1/3, or 2/9. Similarly, the probability that the score is love–fifteen and that John then wins a point is 1/3 times 2/3 which is also equal to 2/9. It follows that the probability of a fifteen all score is 2/9 plus 2/9, or 4/9.

If you have not had a course in probability, convince yourself that these calculations are correct. When should one multiply probabilities? When should one add them?

Continuing in this way, we can calculate probabilities for every score in Figure 5.2.

If we then add the probabilities for the outcomes W_1, W_2, and W_3, what will we have?

Not the probability that John will win a game, but almost that. We still have to add in the probability that the score will reach deuce (40–40) and that John will then win. That leads to a whole new tree, as in Figure 5.3. We leave you to explore how it would work.

FIGURE 5.3
Showing, again from John's viewpoint, possible outcomes in a match after the score has reached deuce. Notice that in theory the game could go on forever, but later stages in the game become more and more improbable.

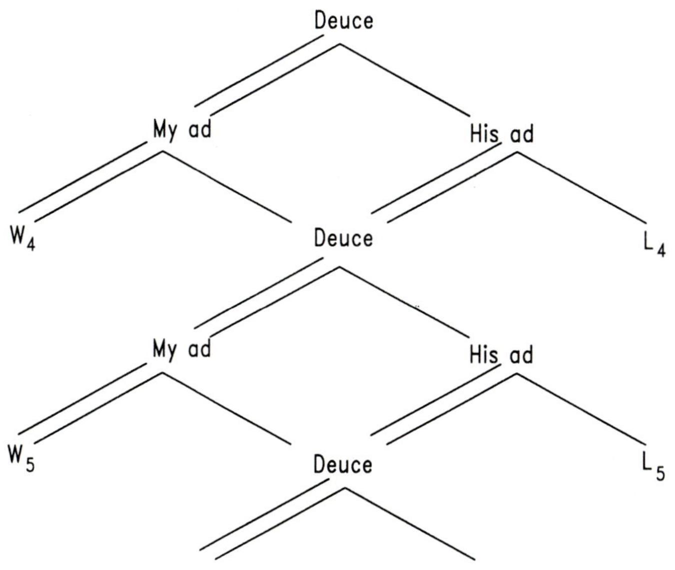

Notice that you could make all the calculations for Figure 5.2 without a computer. However, Figure 5.3 could go on and on, which suggests that you will need a computer to make the calculations.

Or will you? Not if you are clever about it. See if you can show that if the score reaches deuce, the probability that John will go on to win is 4/9 divided by 5/9, or .80.

This approach has a considerable advantage over our simulation model: It calculates the probability of winning a game directly, without the need for a large number of replicates.

Which is the preferable method: the analytic or algebraic approach of Figures 5.2 and 5.3, or the simulation model we built earlier on? Take a few minutes to write down the points for and against each of these approaches.

For a start, the analytic approach is more elegant. It is always satisfying to get a neat mathematical answer instead of having to use a computer. But elegance apart, there are good reasons for taking the analytic approach. One is that you are less likely to make stupid errors that way (such as undetected "bugs" in a computer program). Another is that it is easier to communicate; instead of having to say "This is what I did on the computer" you can write out your solution with all the clarity of a mathematical theorem.

However, the simulation model has one overriding advantage: It is much more flexible. There are small changes one could make to the model that would be impossible (or very difficult) to include in the analytic solution but almost trivial to add to a simulation model. For example, suppose a careful look at the data showed that winning three points consecutively gave John a surge of adrenalin. For the next four points his probability of winning shoots up to .9. Could you incorporate this into an analytic model? Easily?

WHO TO BLAME?

Suppose you decide, on the basis of your modeling, to take the bet. Alas, at the end of the tournament you find that, on balance, you have lost money!

Whom do you blame? Yourself? Your model? Or us?

Unless there is a "bug" in your model, nobody is to blame. Chance is a part of this problem, and calculating risk provides no guarantee (except in the long run) against an unfavorable run of luck. The decision was good; it is the luck that was bad!

SIMILAR PROBLEMS

This problem illustrates two themes:

• The first is that you can calculate the probability of a sequence of events when you know the probability associated with each component event.

• The second relates to the concept of variance: How does the likely outcome of repeated stochastic events depend on the number of replicates?

There are a number of realistic problems that illustrate the same themes. Some can be solved quite easily without a computer. Others can be solved with perseverance or more advanced mathematical skills, but can easily be solved on a computer without any loss of insight into the problem.

Genetic Calculations

A married couple discovers that a combination of their genes is likely (with a probability of .2) to produce a baby with a rare genetic disorder. Ninety-six out of a hundred healthy babies survive to age 21, but a child with this genetic disorder only has a one in three chance of living that long. If the couple have four babies, what is the probability that all four will survive to 21? What is the probability that only three will survive that long?

A Gambler's Ruin

Here is a variation on the famous "gambler's ruin" problem.

A roulette wheel has an equal number of red and black numbers. You are betting on red. If you win, you double your money, i.e., you get back the amount you wagered plus an equal amount. If you lose, your loss is limited to the amount you wagered.

Like so many gamblers, you believe that you can beat the odds if only you have a "system." Your plan of action is as follows: you will bet $5 on the first spin. If you lose, you will keep on doubling your wager (you will bet $10, then $20, then $40 and so on) until you win.

Your objective is to make a profit of $20. So whenever you win, you will start again with a $5 bet and repeat your system. You will leave the table only when you have made your $20 or lost your life's savings (which amount to $315). Nobody has enough faith in your system to lend you any money.

What is the probability that you will lose your savings?

A Different Kind of Gambling

You are a good student, but your performance in examinations depends very much on your mood.

You tend to be in a good mood 4 days out of 6. When you are in a good mood, there is a .9 probability that you will score 100 percent in an examination and only a .1 probability that you will score 80 percent.

When you are in a bad mood there is a .2 probability that you will score 100 percent, a .3 probability of scoring 80 percent, and a .5 probability of scoring only 50 percent.

Your teacher offers you a choice. Either you write two equally weighted examinations (a midquarter and final) or else you write ten equally weighted tests (one each week).

Analyze this proposition and explain your choice.

FURTHER READING

Tools for Thinking and Problem Solving by Moshe Rubinstein (Englewood Cliffs, NJ, Prentice-Hall, 1987) has a strong emphasis on stochastic problems. It also has an interesting (and thorough) discussion of heuristics, decision making, and how to structure decision situations.

In Chapter 1 we mentioned *An Introduction to Models in the Social Sciences* by C. A. Lave and J. G. March (New York, Harper & Row, 1975). It also contains problems in a decision-making context that depend on probabilistic concepts.

Anthony Starfield and Andrew Bleloch develop and describe a stochastic model of the population dynamics of a rare antelope in the third chapter of their book: *Building Models for Conservation and Wildlife Management* (originally published in New York by Macmillan, in 1986, but now marketed by McGraw-Hill). This chapter pays particular attention to how you go about drawing conclusions from stochastic models.

FOOD FOR THOUGHT

THE PROBLEM

You are planning a dinner party for yourself and five guests.
This is your menu:

Onion soup with French bread
Roast chicken with boiled mixed vegetables
Pie with hot chocolate sauce

You have arranged to go to a baseball game before the dinner, and afterward the six of you are going to the theater. You can prepare the ingredients for the meal before you leave for the match, but will you be able to get back in time to cook and serve the dinner? Will you and your guests get to the theater on time?

Your problem is to estimate the *shortest* time in which you could cook, serve, and eat the meal. (The clearing and washing can wait until the next morning.)

THE DATA

The following are your best estimates of the times needed to complete the various activities associated with the dinner:

Cooking the onion soup	30 minutes
Warming the bread in the oven	10 minutes
Roasting the chicken in the oven	45 minutes
Boiling the vegetables	25 minutes
Warming the pie in the oven	15 minutes

Cooking the chocolate sauce	40 minutes
Serving and eating the soup	20 minutes
Serving and eating the chicken	25 minutes
Serving and eating the apple pie	15 minutes

You have been saving two bottles of a good red wine for an occasion such as this. You need to allow 5 minutes to uncork the wines and 30 minutes to let them breathe. You intend to serve the wine with the chicken, even though you know your friend Andrew will pass a snide remark about serving red wine with poultry.

Incidentally, you have only one pot to use for boiling the soup, the mixed vegetables, and the chocolate sauce. (You can assume that it takes no time at all to rinse it.) Also, your oven can hold only one item at a time.

How quickly can you cook, serve, and consume the meal?

You should be able to get the answer in 15 to 20 minutes, but take a little longer to think about how you got there. How, for instance, would you persuade somebody that the answer you have found really is the shortest time?

✔

YOUR SOLUTION

What is your answer?

Is it less than 110 minutes? If so, are you sure you have not forgotten something?

Is it more than 110 minutes? If so, what assumptions did you make? Perhaps you insisted on serving hot food directly from the oven. We assumed that we had a warming tray (or that our friends would not complain if the bread was not all that hot when we served it with the soup).

More importantly, how did you get your answer? Did you develop an organized way of representing the data? Or did you just somehow muddle through? Can you reproduce and explain your calculation in a convincing way?

There are a host of different ways in which you might have represented the data in an organized fashion.

Perhaps you took a piece of paper and divided it into a number of columns, as in Figure 6.1. In the first column you might have jotted down the time in 10-minute intervals. The second column could have shown what was cooking in the pot, the third what was in the oven, and so on. With a pencil and eraser

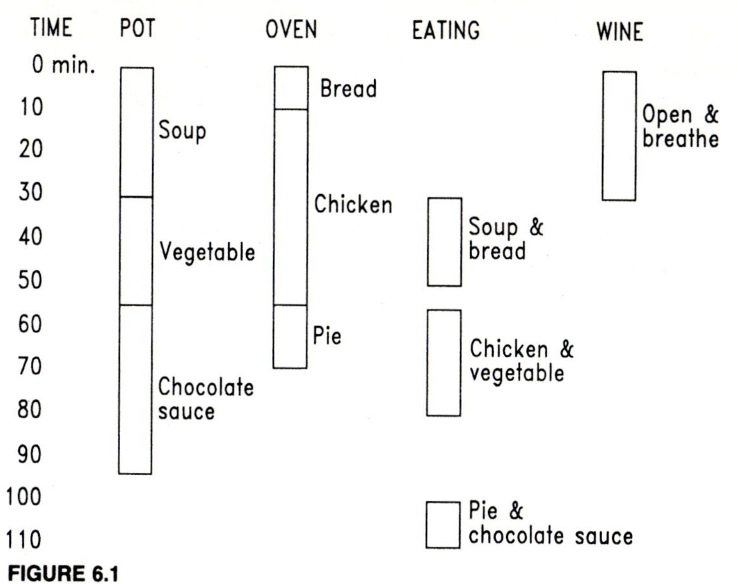

FIGURE 6.1
This representation of the dinner scheduling problem highlights the use of the pot and oven.

you could easily have represented alternative schedules and figured out how to complete it all in as short a time as possible.

Alternatively, you might have taken a sheet of graph paper, as in Figure 6.2. Here the horizontal axis represents time. There is no vertical axis, but each activity (such as cooking the soup) is represented by a bar parallel to the time

FIGURE 6.2
Another representation of the dinner scheduling problem, showing all the activities without any direct reference to resources such as the pot and oven.

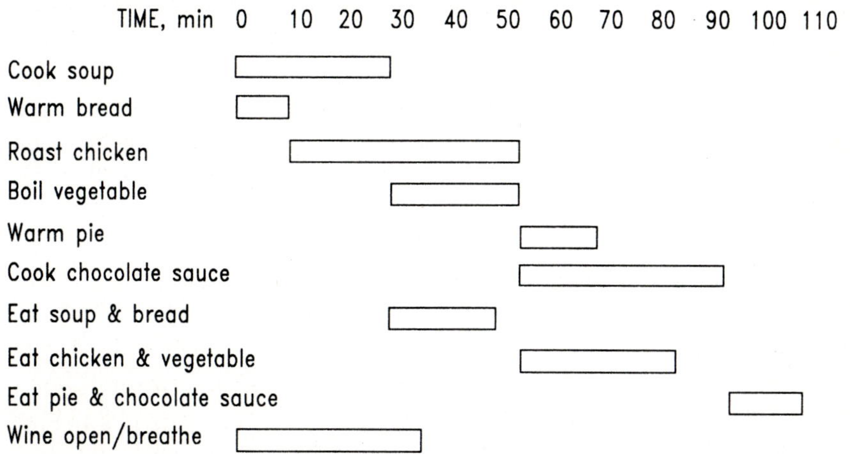

axis. The length of the bar represents the duration of the activity (30 minutes in the case of cooking the soup) and its position shows precisely when it starts and stops. Again, with a pencil and eraser, you could have rearranged the order of the bars and solved the problem in about 10 minutes. Figure 6.2 is, for obvious reasons, called a *bar chart*.

Did you develop an approach that was similar to either of these? Notice that they are essentially the same representation. The only difference is that one is tabular while the other is graphical. Both enable you to represent the data you have been given in a form that makes it easier to manipulate.

Looking at these two representations or your own, what would you say are the key concepts that characterize this kind of problem? Go ahead and make a list of them. If you are unsure of what we mean by "key concepts," think of the key words that you would use to describe this problem and the ideas you need to consider when you try to solve it.

KEY CONCEPTS

Our list looks as follows:

- *Time:* Imagine that a stopwatch is clicked on when you walk into the kitchen and clicked off when the last guest has eaten the last crumb of pie.
- *Activities:* Each activity has a starting time, a duration, and an ending time.
- *Interdependence:* This would be a trivial problem if one did not have to consider how a suitable time slot for one activity depended on when other activities were scheduled. There are two aspects to this interdependence: concurrence and precedence.
- *Concurrence:* Some activities can occur simultaneously, others cannot. For example, the soup could be boiling while the bread is warming in the oven, but the vegetables cannot be cooked while the pot is being used for the soup.
- *Precedence:* Certain activities cannot start before others have ended. For example, the soup should be cooked *before* the vegetables, and the vegetables cannot be boiled until the soup is ready. The order is important.
- *Early starting time:* Taking precedence into account, one can ask how early each activity could possibly start. For example, the earliest time to put the soup on is right at the beginning (i.e., at zero minutes). Since the soup takes 30 minutes to cook, the early starting time for the vegetables will be at 30 minutes. Early starting time is an important concept because if *all* activities are scheduled at their early starting time, then we are *guaranteed* that the dinner will be completed in the shortest possible time.

How does this compare with your list? You may have used different words, but did you have similar ideas? Do you agree that these are the key concepts? Perhaps you thought time was so obvious that you left it off your list. It is always better to be explicit, even about the obvious.

How well do the representations in Figures 6.1 and 6.2 reflect the key concepts?

✓

Both figures do a good job of representing time and activities. Figure 6.1 goes some way toward making interdependence explicit because the columns have been chosen intelligently: they remind you that there is only one pot and that only one item can be placed in the oven at a time. Both figures also show precedence, but only after you have done the work of sorting out explicitly which activities must be completed before another begins. Neither figure *helps* you to do this. You are left to draw on information from elsewhere (the statement of the problem), and you have to bear that information in mind while you "juggle" with your pencil and eraser.

In the same vein, both figures clearly show the early starting time for each activity (provided you have drawn them with each activity starting as early as possible), but neither helps you to actually calculate early starting times.

Both representations are adequate for solving this problem. But how well do you think they would work if you had a similar problem with four or five times the number of interdependent activities? Would it still be possible to juggle effectively?

How well would your method of solution have worked on a much more complex problem?

If both these representations are strong on time, but weak on interdependence, can you think of a representation that is strong (i.e., explicit) about interdependence?

Try to draw a diagram that clearly tells you which activities can be occurring simultaneously and which must be completed before others can begin. Your representation may already do this, in which case read on. Otherwise see if you can modify it, or try to think of a new representation.

✔

PRECEDENCE DIAGRAMS

Figure 6.3 is one way of representing precedence explicitly. Here each activity is represented by an arrow or directed arc. At the head and tail of each arrow (or arc) is a node. *Nodes* represent the beginning and/or end of an activity. For example, the node marked 3 in Figure 6.3 represents the end of the soup cooking activity and the beginning of the vegetable cooking activity.

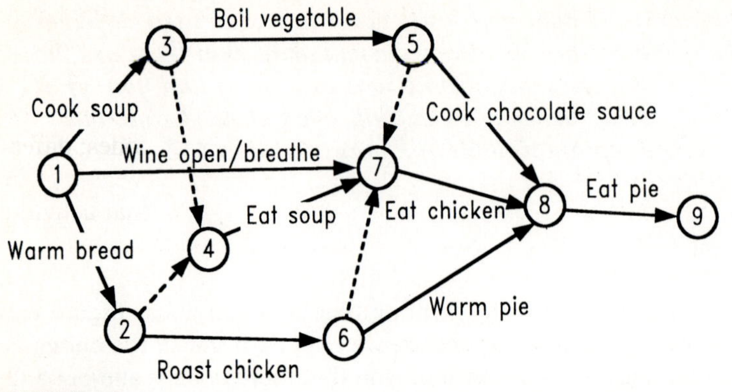

FIGURE 6.3
A representation of the dinner scheduling problem that explicitly shows the interdependence of the activities. It is called a *precedence diagram.*

No two arcs are permitted to start and end (as in Figure 6.4) on exactly the same two nodes. Dotted arrows or arcs are used to avoid this or to clarify links between the precedence paths; you can think of them as "dummy" activities with zero duration.

There are other ways of representing precedence. For example, one could draw a diagram where the nodes represent activities. (What would the arrows represent then? Interdependence?) However, Figure 6.3 is a particularly useful representation for our purposes, as you will see.

Notice that we have numbered all the nodes in Figure 6.3.

What is special about the order of our numbers?

Right, the starting node is numbered one and the end node has the highest number. But there is more to it than that! Notice that we have chosen the num-

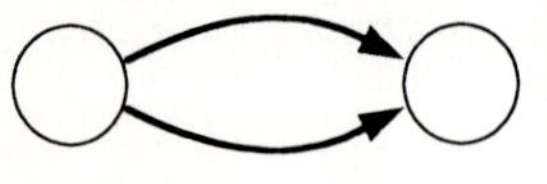

FIGURE 6.4
If two activities start and end on the same node (top diagram) then a dummy activity must be introduced to make the path between any two nodes unique (bottom diagram).

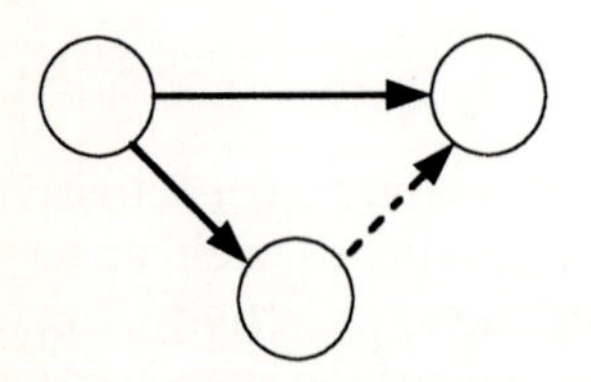

bers in such a way that the node at the beginning of an activity always has a lower number than the node at the end of that activity. This holds even for dotted arrows.

It is very easy, once you have drawn the precedence diagram, to number the nodes in this way. There is no unique match between numbers and nodes; different patterns of numbers can satisfy the condition that the node at the beginning of an activity must have a lower number than the node at the end of that activity. With a little practice you will develop your own style of numbering.

Why do you think this is useful?

Try redrawing the bar chart of Figure 6.2 with the help of the precedence diagram. Write the node numbers at the beginning and end of each bar.

✓

Figure 6.5 has been drawn in this way.

Notice that using the node numbers makes it much easier to draw the bar chart. The numbers have the effect of building precedence into it.

The combination of precedence diagram (numbered in this special way) and bar chart facilitates the solution of the problem. Together, they provide a

FIGURE 6.5
Incorporating the activity numbering of Figure 6.3 into the time representation of Figure. 6.2.

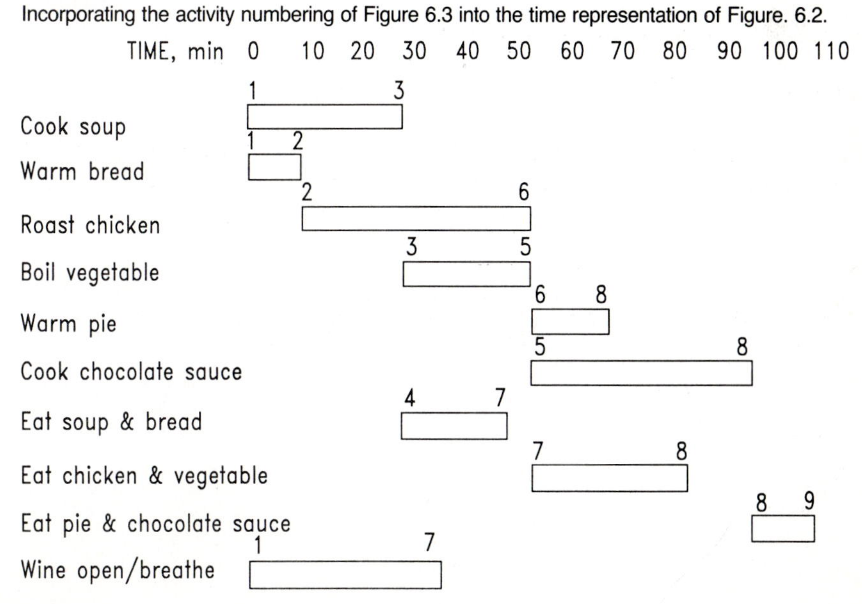

methodology that enables you to solve even much larger problems accurately and efficiently.

Alternatively, the representation in Figure 6.3 can be used without a bar chart for making time calculations. In particular it lends itself to calculating early starting times.

Can you think of a way of introducing time into Figure 6.3?

We can associate time with the nodes. In particular, we are interested in the early starting time. The following algorithm can be used to construct the information contained in Figure 6.6. Here we have calculated and written, next to each node, the early starting time for the activities that start at that node:

1. First write the duration of each activity on the arrow that represents it.
2. Write a zero (for time equals zero minutes) next to the starting node (node 1).
3. Then work your way through the nodes *in order* (i.e., to node 2, then node 3, and so on). At each node do the following:
   a. Consider all the arrows ending on that node. These are the activities that must be completed before any of the activities leaving the node can begin.
   b. For each arrow, take the early starting time at the beginning of the arrow and add that to its duration.
   c. The *largest* time calculated in this way must be the early starting time for the node. Write this next to the node.

FIGURE 6.6
The precedence diagram of Figure 6.3 again. Now it shows the duration of each activity, making it easy to calculate early starting times. These are shown above each node, followed by a slash /.

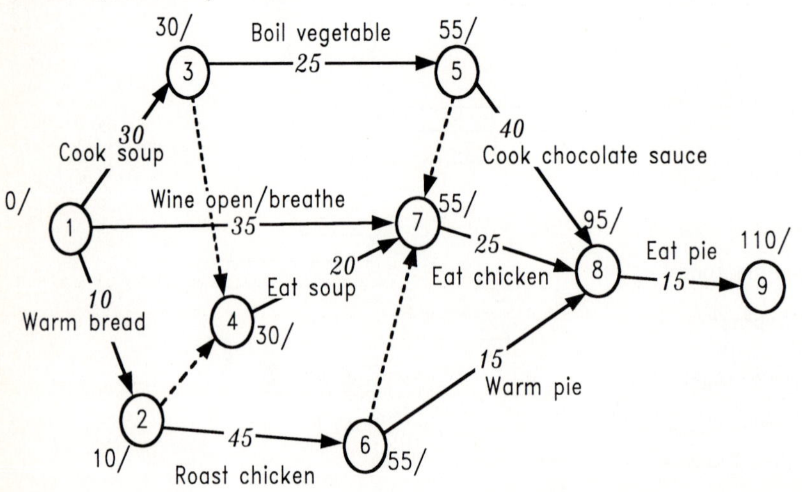

For example, suppose you have already calculated the early starting times for nodes 5, 6, and 7 in Figure 6.6. If you look at node 8, there are three arrows that end on it: 5-8, 6-8, and 7-8. The duration of 5-8 is 40 minutes, and the early starting time at node 5 is 55; adding 40 to 55 gives **95** minutes. Similarly, the duration of 6-8 is 15 minutes, and the early starting time at node 6 is also 55; adding 15 to 55 gives **70**. Finally, for activity 7-8 we add 25 to 55 (again) to get **80** minutes. Comparing these three times (95, 70, and 80), you see that the largest is 95, and this is what you write next to node 8. It is the early starting time for activity 8-9.

Notice the payoff from our simple system for numbering the nodes. It leads to a foolproof algorithm for calculating the early starting times. (Foolproof in the sense that it reduces the likelihood that you will leave out an alternative, perhaps longer, pathway to the node you are considering.) If you do not appreciate this, try numbering the nodes at random, and then try to calculate early starting times!

Finally, notice that when you get to the last node (after eating the pie) the early starting time written next to it is in fact the answer to your problem!

SENSITIVITY AND TRADE-OFFS

Your dinner guests will be late for the theater; 110 minutes is too long! You want to reduce time without, if possible, reducing the quality of the meal. Three options spring to mind:

1. Serve a wine that does not need to breathe.
2. Leave out the French bread.
3. Borrow an extra cooking pot.

Investigate these three options.

How much time would each save? Could you save more time by exercising two of the options instead of one? Or all three?

✔

How did you answer the above questions? Was your own representation helpful? Or did you find one of our representations (Figure 6.1, 6.5, or 6.6) useful?

Do you have anything to add to your list of key concepts as a result of trying to answer the above questions?

✓

We found Figure 6.5 helpful. We first looked at activity 1-6 (uncorking the wine and letting it breathe). The activity starts at time zero, but if we imagine shifting the bar to the right along the time axis, we see that it would make no difference whatsoever if we forgot to uncork the wine and remembered it only after 10 or 15 minutes, or even longer.

What can we conclude from this?

If delaying the wine makes no difference to our answer, then we conclude that there is nothing to be gained by substituting a wine that does not need to breathe.

That answers our immediate question, but we were intrigued by the idea of moving the activity bar along the time axis. We decided to explore the question: "At what stage would it be too late to uncork the wine?" Digress for a moment and see if you can answer that question.

✓

Is your answer 20 minutes? Or is it 35 minutes? Which is the correct answer?

If you shifted activity 1-7 until the end of that activity coincided with the beginning of activity 7-8 (serving and eating the chicken) then you could have delayed 20 minutes before opening the wine. But notice that you could also have delayed serving the chicken by 15 minutes! So, in fact, you could have opened the wine after 35 minutes and still completed the meal in the original 110 minutes.

This digression leads us to two new concepts. We already have the concept of an *early starting time*. If we think about how far to the right we can shift an activity (without prolonging the meal), we also have the concept of a *late finishing time*. The late finishing time for activity 1-7 is 70 minutes; for activity 7-8 it is 95 minutes.

Suppose we calculate the difference between the late finishing time and the early starting time. The answer must surely be greater than or equal to the duration of the activity. This leads to the second concept. We define the *slack* of an activity as follows:

$$\text{Slack} = \text{late finishing time} - \text{early starting time} - \text{duration}$$

We can also formulate a rule: If the slack of an activity is greater than zero, then you cannot reduce the total time for the dinner by eliminating or speeding up that activity by itself.

Now look again at the alternatives of leaving out the French bread or borrowing an extra pot. Do our new concepts help to evaluate these options? Is the rule we have just promulgated useful?

✓

From Figure 6.5 we see that activity 1-2 (warming the French bread) has positive slack. We therefore gain nothing by leaving out the bread.

On the other hand, all the activities that use a pot (cooking the soup, the vegetables, and the sauce) have zero slack. So it does make sense to borrow an extra pot. We could use it for the vegetables, to begin with, and then rinse it and use it for the chocolate sauce.

How much time would that save?

The answer is 15 minutes.

Did you find it by looking at Figure 6.5, or did you draw a new precedence diagram and bar chart?

You might have been able to get the answer without drawing new diagrams, but if you are going to explore possibilities more fully, you will find it helpful to at least redraw the precedence diagram and calculate new early starting times. You can then ask yourself questions such as: "If I had two pots, would it then make sense to change the wine and/or leave out the bread?"

We leave you to answer such questions. We also leave you to investigate alternative ways of saving time.

CRITICAL PATHS

In the previous section we used bar charts to explore different ways of reducing the total time for the meal. Similar conclusions can be drawn from the precedence diagram.

In Figure 6.6 we calculated early starting times systematically at each node in the precedence diagram. In Figure 6.7 we also calculate and include late finishing times.

These are calculated systematically too, this time working *backward* from the highest-numbered node down to node 1 at the beginning of the diagram.

At each node we look at all the activities that begin at the node. For each activity we subtract its duration from its late finishing time. The *lowest* number we get in this way is the late finishing time for all the activities that end at that node. We write it down next to the node, separated from the early starting time by a slash.

How could we calculate slack from Figure 6.7?

Consider, for example, activity 4-7 (serving and eating the soup). The late finishing time (at node 7) is 70 minutes and the early starting time (at node 4) is 30 minutes. The duration is 20 minutes. It follows that the slack is

$$70 - 30 - 20 = 20 \text{ minutes}$$

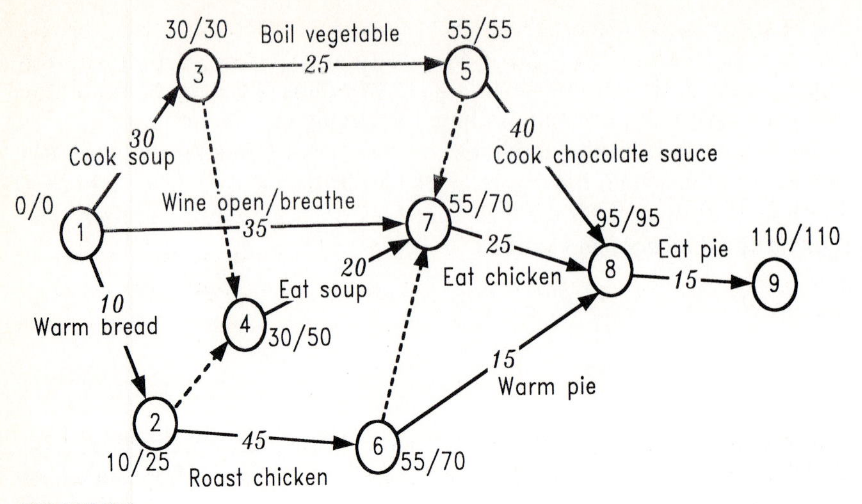

FIGURE 6.7
Late finishing times are included in this version of the precedence diagram. The notation is early starting time/late finishing time.

Similarly, the slack for activity 7-8 (serving and eating the chicken) is

$$95 - 55 - 25 = 15 \text{ minutes}$$

Does this mean that we could dawdle for an extra 20 minutes over the soup and for an extra 15 minutes over the chicken without being late for the theater?

Not so! Slack is a more subtle concept. It is the amount of time you can spare provided you don't waste time anywhere else. Using some of the slack for one activity can eat (no pun intended) into the slack for another activity.

Can you identify those activities in Figure 6.7 that have zero slack? How do you quickly recognize activities with zero slack?

Consider activity 5-8, for instance (cooking the chocolate sauce). The numbers next to node 5 are both 55 and the numbers next to node 8 are both 95. Moreover the difference between them is 40, which is exactly the duration of the activity.

All these conditions must be satisfied for zero slack.

Notice that the activities with zero slack in Figure 6.7 lie along the path that passes through nodes 1, 3, 5, 8, and 9. We call this the *critical path*.

How does it help to know whether or not an activity lies on the critical path?

In two ways. First, those are the activities you should look at whenever you want to reduce the total time. Second, you know that if *any* activity on the critical path is either delayed or takes longer than expected, then the total time must be increased.

A DIFFERENT REPRESENTATION

What have you learned from this chapter? Take a few minutes to list the most significant or the most interesting ideas that you have learned.

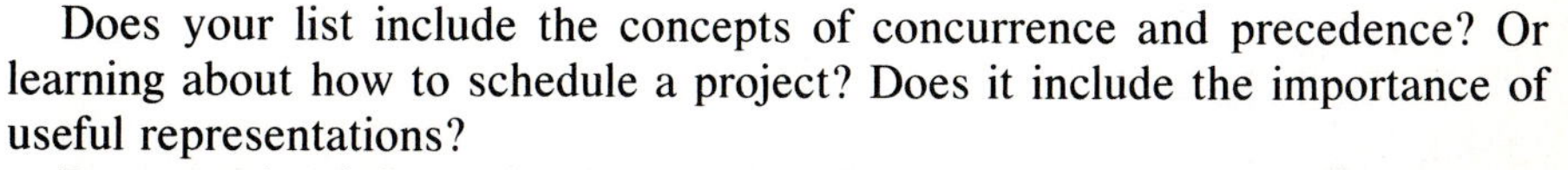

Does your list include the concepts of concurrence and precedence? Or learning about how to schedule a project? Does it include the importance of useful representations?

Representing information in an organized way can be a powerful tool. In previous chapters we have seen how important it is to choose the right variables and a sensible notation. In this problem it was important to choose the right sort of representation and a sensible system for numbering the nodes. Notice how one representation rather than another makes it easier to answer all sorts of questions (not just the original question).

Notice too how the node numbering rule made it possible to calculate early starting and late finishing times so systematically that we could program a computer to make the calculations for us.

Think about how you would design a computer program to analyze more complex scheduling problems. In particular, think about what the computer could do and what you would still do yourself. How would you "describe" the problem to the computer?

Did you decide that <u>you</u> would have to draw the entire network yourself before going to the computer? Did you also decide that you would have to number the nodes sequentially before asking the computer to perform the analysis?

Did this bother you? After all, your reason for going to the computer, particularly with larger and more complex problems, is to permit quick and easy updating as delays and other problems occur. It does not make much sense if you still have to cope manually with the changes and the complexity of the problem.

When we tried to implement the critical path algorithm on a computer we found the activity-on-arc representation difficult to work with. We were hop-

ACTIVITIES	DURATION, min	ACTIVITIES THAT PRECEDE
1. Cook soup	30	
2. Warm bread	10	
3. Roast chicken	45	2
4. Boil vegetable	25	1
5. Warm pie	15	3
6. Cook chocolate sauce	40	4
7. Eat soup	20	1 2
8. Eat chicken	25	3 4 7 10
9. Eat pie	15	5 6 8
10. Wine open/breathe	35	

FIGURE 6.8
An *activity-on-node* representation makes it easy to describe a scheduling problem to a computer program.

ing the computer would help us to number the nodes and draw the precedence diagram but found that it was easier to do that with pencil and paper beforehand.

Remember the earlier discussion of what would happen if activities were represented on nodes? Well, the activity-on-node representation is much easier to implement on the computer. The user only has to enter each activity, its duration and its predecessors as shown in Figure 6.8. The computer can then do the rest. The calculation of early start time, late finishing time, and the critical path remains the same.

An activity-on-node representation is shown in Figure 6.9.

What do the arcs represent?

Right! They represent precedence. The critical path calculations for this representation are shown on Figure 6.10.

Use Figure 6.10 to look for ways of reducing the time it takes to serve and eat the meal. You should find it as easy, if not easier, to explore consequences with the activity-on-node rather than the activity-on-arc representation.

DISCUSSION

"This has been an interesting exercise, but what has it got to do with modeling?" Is that what you are thinking? Can you anticipate our reply?

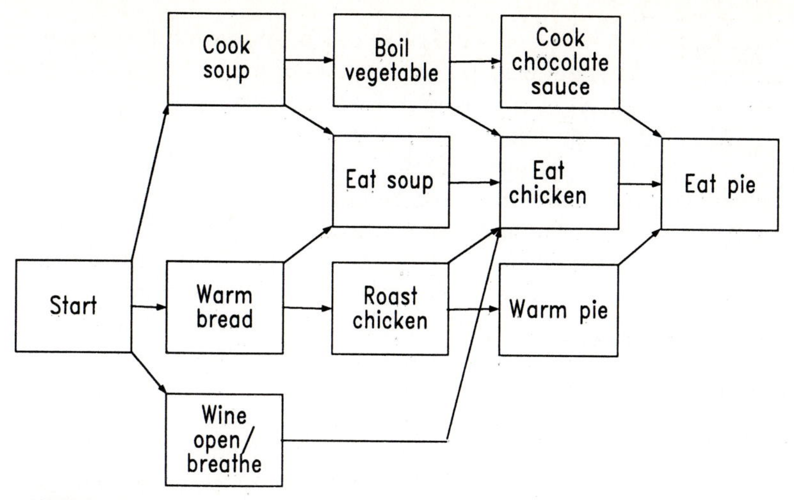

FIGURE 6.9
Activities represented on nodes rather than on arcs. Now the arcs represent interdependence.

FIGURE 6.10
Results of the critical path analysis. All the information in this computer output can be calculated directly from Figure 6.7

ACTIVITIES	DURATION, min	EARLY Start	EARLY End	LATE Start	LATE End	FLOAT
1. Cook soup	30	0	30	0	30	0
2. Warm bread	10	0	10	15	25	15
3. Roast chicken	45	10	55	25	70	15
4. Boil vegetable	25	30	55	30	55	0
5. Warm pie	15	55	70	80	95	25
6. Cook chocolate sauce	40	55	95	55	95	0
7. Eat soup	20	30	50	50	70	20
8. Eat chicken	25	55	80	70	95	15
9. Eat pie	15	95	110	95	110	0
10. Wine open/ breathe	35	0	35	35	70	35

Recall that in Chapter 1 we defined a model as a "purposeful representation." This chapter is all about purposeful representations. A model does not have to lead to a complex set of calculations. The representations we have developed here all lead to simple calculations, but they are nevertheless effective for organizing and manipulating information.

When a scheduling problem is represented in this way, either on paper or through using a computer program, it is much easier to ask (and answer) questions such as "What happens if I make this modification?" or "How could I possibly achieve that goal?" Modelers sometimes refer to these questions as *scenarios*. Making changes, asking questions, and solving different versions of the problem can lead to further understanding of the problem's environment. The goal of modeling is to develop that understanding.

Remember, the ideas are what is important. In 1961, when people were just beginning to use computers for problem solving, R. W. Hamming (*Numerical Methods for Scientists and Engineers,* Tokyo, McGraw-Hill, 1962) wrote: "The purpose of computing is insight, not numbers!" The same should be said for modeling.

In the process of solving this problem we have met a standard technique, generally known as *project scheduling by the critical path algorithm*. We did not intend to directly teach you the technique; our objective was to lead you toward developing it yourself, or at least toward appreciating that there was a need for a technique like this. That, in practice, is how (and why) techniques are "developed"; they are a clever response to a class of problems rather than something one learns by rote from a textbook.

What heuristics have we used in this chapter?

We have certainly made good use of the heuristic that suggests we try to represent our problem in different ways. We have also used modified versions of heuristics that we met in earlier chapters. For example, we tried to identify key concepts rather than the important variables, and we looked for a node numbering system rather than a suitable notation. The thought processes are, however, still the same. They led to an *algorithm* in this case rather than a formula or set of equations.

What have we forgotten to do in this chapter? What are the hidden assumptions in the way we have tackled this problem?

This is always a good question to ask. You might be able to think of a host of assumptions we have made. We recognized only one major assumption: we have assumed that the data given at the beginning of this chapter are accurate.

What happens if it takes 50 instead of 45 minutes to cook the chicken?

Should we build a different sort of model that takes into account the uncertainty in the data? Should we have built a stochastic model?

Our answer to this would be "no." Of course the data are inexact, and of course it makes a difference to our answer when we change some of the numbers, but we would argue that the model we have built enables us to explore the question of uncertainty in the data in exactly the same way that we have explored other questions (such as "Will we save time if we leave out the wine?"). Uncertainty in the data is just another what-if situation. Uncertainty does not *overshadow* the problem to such an extent that we need to build it into our representations.

We leave you to think about what you would do if uncertainty were a major consideration in a scheduling problem.

FURTHER READING AND SIMILAR PROBLEMS

Suppose that instead of organizing a dinner party you were project manager for the construction of a skyscraper. How would you schedule and keep control over all the activities taking place on the building site? Think about how precedence diagrams and bar charts would help you.

Suppose too that you were sole contractor for the entire construction job. Think about the workers you would have to hire at various stages of the project and those you would have to lay off as tasks are completed. In practice you would be looking for workers with different skills at different stages in the project (an electrician does not install windows), but suppose for the moment that your local union is rather unusual: Each member can perform any task on a construction site. You now have an added dimension to your problem: You want the construction completed on time, but you also want the size of your labor force kept relatively constant. You do not want to hire and fire too often, and you do not want to pay workers who have nothing to do. Think about how you would represent your work force on a diagram and how you might use such a diagram in conjunction with your project bar chart to minimize repeated hiring and firing.

The above exercise is an example of what is known as *resource leveling*. You can read about critical path methods, resource leveling, and other exciting techniques (such as *crashing*) in the chapter on project scheduling in *Design and Planning of Engineering Systems* by D. D. Meredith, K. W. Wong, R. W. Woodhead, and R. H. Wortman (Englewood Cliffs, NJ, Prentice-Hall, 1985). Incidentally, this book also covers problem formulation, systems modeling, decision analysis, systems simulation, and the engineering process.

However, the main purpose of this chapter was not to introduce you to project scheduling. Rather, we wanted to show how helpful it is to find an appropriate representation when you have to deal with a number of interlinked events or processes. Here are two problems that one would not necessarily solve by the critical path method, but where representation is the crucial part of the problem:

1. You are the owner of a small airline company. You have just acquired another small company. The time has come to choose a base for your operations. Given information about the cities you serve and passenger volume, which airport should you choose as your "hub"?
2. You are a traveling sales representative. You know which cities you have to visit during the next 2 weeks. Develop a representation that will help you organize your itinerary.

Finally, if you have already looked at one of the books we recommended in the previous chapter (*Tools for Thinking and Problem Solving* by Moshe Rubinstein), you will have noticed that it contains a number of interesting and useful tools for representations. If you have not looked at the book yet, perhaps now is the time to see what you can find about representations in it.

THE STUDENT'S DILEMMA: FRENCH, CALCULUS, TIME, AND MONEY

BACKGROUND

You will be taking your final examinations in French and Calculus in 10 days' time, but, instead of preparing for these examinations, you have been having much more fun learning to model on your personal computer.

Unfortunately, your advisor seems to be more interested in your academic performance than your intellectual development. You have just had a rather unpleasant meeting with her. She has told you, in no uncertain terms, what will happen to you if you fail to get a passing grade in either examination. In fact, you need a marked improvement in your overall performance to keep out of trouble.

This meeting worries you. You have to admit that you are not prepared for the examinations. If you don't make the best possible use of the time available, you could easily fail.

What you need is some concentrated tutoring. There is a language laboratory that has some excellent review tapes that would be ideal preparation for the French examination. The only problem is that it costs $5 per hour to use the laboratory.

Preparing for the Calculus examination is going to be more difficult. Your friend recommends a good tutor, but demand has pushed up prices: The tutor charges $15 per hour.

Your budget will stretch to at most $100 for these extra activities.

You have always had attention span problems. You cannot imagine spending more than 8 hours at the tapes during the next 10 days. You could also not absorb more than 7 hours of Calculus tutoring in 10 days. But that does not mean you are willing to spend a total of 15 hours on these activities. When you look at your social calendar for the next 10 days, you find that, at most, you could squeeze in 11 hours of extra activity, be it tutoring or time in the language laboratory, or a combination of the two.

What is the best way to spend your time and money?

This question is a wonderful excuse for getting back to modeling. You can even feel virtuous about your impending examinations while you try to answer it!

YOUR FIRST TASK

Your first task is to make sure you understand the problem.

Organize the information you have been given in a clear, unambiguous way. Define the objectives of the exercise. Decide what additional information, if any, you require and make a list of questions you wish to ask and/or assumptions you would like to make.

Since your time is at a premium, do not spend more than half an hour on this task.

✔

OBJECTIVES, QUESTIONS, AND ASSUMPTIONS

How many assumptions are on your list? What are they?

More important, how did you define your objectives?

Perhaps you defined the objective as follows: To allocate time to the two subjects in such a way as to score the highest possible aggregate mark, without failing either examination.

If this (or something similar) was your objective, did you recognize that the time allocation would have to fit

- *Your social calendar?*
- *Your attention span?*
- *Your budget?*

Where did you need additional information?

Did you see the need to relate the time spent at the language laboratory to your likely performance in the French examination? Also the time spent with a tutor to performance in the Calculus examination?

If so, you probably asked yourself the following questions (or put them on your list of questions that would have to be answered):

1. "How well do I think I could score in a French examination without any further preparation?"

2. "For each hour spent in the language laboratory, how many extra marks am I likely to score?"

3. "How well do I think I could score in Calculus without any additional tutoring?"

4. "How many extra calculus marks am I likely to achieve for each hour of tutoring?"

Also, how are quizzes scored in both classes? What is a passing grade? Are French and Calculus equally important to your advisor?

Assumptions

Let us suppose that all quizzes in your college are graded out of 100. A passing grade is 50. Your advisor will first look at your grades separately (to make sure that you have passed in both subjects) and will then look at the sum of the two grades (as a measure of your overall performance).

From previous experience you expect that you could score about 40 percent in each examination if you studied in your usual relaxed fashion, without tutoring or time in the laboratory.

How will the laboratory help you? You have used it before—occasionally. You reckon that 3 intensive hours in the laboratory should boost your French grade to a bare pass.

You have had no previous experience with the Calculus tutor, but after discussions with your friend you are confident that you could score 50 percent in Calculus after only 2 hours of tutoring.

Are there any additional assumptions you need to make?

We could not think of any at this stage.

NOW FIND A SOLUTION TO YOUR PROBLEM

Now that you have agreed on assumptions and objectives, go ahead and solve the problem. How are you going to allocate your time?

HOW DID YOU GET YOUR ANSWER?

As you may have guessed, your answer is not as interesting to us as the way you went about solving the problem.

Quickly jot down the steps that led to your solution.

Testing All Plausible Combinations

Perhaps you argued that since there were only a limited number of combinations, you could draw up a table such as Table 7.1 that tested all of them. Notice that the table has six columns:

1. Hours in the language laboratory
2. Hours of tutoring
3. Total cost
4. Likely French grade
5. Likely Calculus grade
6. Total grade

Notice too how many combinations are impractical, either because of your attention span problem or because they are too costly or because you are still likely to fail Calculus. In the end there are only four or five sensible combinations, and the one that leads to the highest total grade is 7 hours in the language laboratory and 4 hours of tutoring.

You like this solution; it leaves you with $5 in your pocket.

But what additional assumptions are implicit in the above approach? There are at least two new assumptions built into Table 7.1. Can you identify them?

First, the table assumes that the time spent in the language laboratory and with the calculus tutor must add up to 11 hours, if at all possible. This may be a good assumption, but perhaps it is may not. You might, for example, want to spend so much of your budget on Calculus that you cannot afford to spend the rest of your time in the language laboratory.

Second, the table assumes that the smallest time unit, in both the laboratory

TABLE 7.1
PLAUSIBLE COMBINATIONS OF HOURS IN THE LANGUAGE LABORATORY AND HOURS OF CALCULUS TUTORING

Language lab, h	Tutoring, h	Cost, $	Expected French grade	Expected Calculus grade	Expected total grade
8	3	85	66.7	55	121.7
7	4	95	63.3	60	123.3
6	5	Too expensive			
6	4	90	60	60	120
5	6	Too expensive			
5	5	100	56.7	65	121.7
4	7	Too expensive			
4	6	Too expensive			
4	5	95	53.3	65	118.3
3	7	Too expensive			
3	6	Too expensive			
2	6	100	Fail		

and tutorial, is an hour. Again, this may be a good assumption, but for the sake of argument let us suppose that both the tutor and laboratory have a schedule based on 15-minute intervals and that both charge accordingly. What are you going to do then?

Sure, you could still construct a table of combinations, but notice that the number of combinations would be that much larger.

Solution by Argument

Alternatively, you might have tried to argue out the solution. Your argument might have gone as follows:

"Money is the real problem, so where do I get the best value for my dollars? The language laboratory only charges $5 per hour, and I am likely to score 3⅓ extra points for each hour, so the cost is $1.50 per point. The cost per Calculus point, on the other hand, is 15/5 which is $3 per point. Therefore I should spend as much time as possible in the language laboratory.

"The most time I can handle in that laboratory is 8 hours, which leaves 3 hours free for tutoring. Yes, I could pass the Calculus quiz!

"My total cost would only be $85 (it would be nice to save $15!) and my grades are likely to be 55 for Calculus and 66.7 for French, giving me a total of 121.7."

But notice Table 7.1, for all its assumptions, led to a better solution: a total of 123.3 points! What is the flaw in the above argument?

✓

The argument began by assuming that money was the problem, but ended with a solution that only cost $85. It follows that money was not necessarily the real problem.

"Right!," you say, "So let us assume that time is the problem. In that case my time is better spent with the tutor: I earn 5 points per hour there as compared with 3⅓ points per hour in the language laboratory.

"I need at least 3 hours in the language laboratory if I am going to pass the French quiz, so I will spend my remaining 8 hours with the tutor. My final grades are likely to be 50 for French and 80 (Wow!) for Calculus, giving a grand total of 130 points.

"My total cost will be $15 plus $120—*I am over my budget*! So, you see, money was a problem after all!

"Perhaps I should replace tutor time with language laboratory time until I am within budget. For each hour I replace in this way, I save $10, so I need to replace 3½ hours. That would give me 4½ hours of tutoring and 6½ hours of French. I'm just checking.... Yes, that will cost exactly $100, and my expected grades are 62.5 for Calculus and 61.7 for French, giving a total grade of 124.2. That really is the best answer so far!"

And notice that you made fewer assumptions this time.

A Mathematical Formulation

Alternatively, you might have tried to find the answer algebraically. Perhaps you felt that there were so many different aspects to this problem that it might help to write them all down in a succinct mathematical way.

How many variables do you have? Suggest a notation.

Let x be the time (in hours) spent in the language laboratory; let y be the time spent with the calculus tutor.

Why two variables? Why not let the one be x and the other $(11 - x)$?

The answer is that you know that the total time is less than or equal to 11 hours. You could write this, not as an equation but as an *inequality:*

$$x + y \leq 11 \qquad (7.1)$$

Can you write the other information you have in terms of x and y too?

Yes! The upper limits to the time spent in the language laboratory and with the tutor can be written

$$x \leq 8 \qquad (7.2)$$

and

$$y \leq 7 \qquad (7.3)$$

What about the budget limitation of \$100?

This leads to the inequality

$$5x + 15y \leq 100 \qquad (7.4)$$

And the condition that you have to get a passing grade in both subjects?

How about

$$x \geq 3 \qquad (7.5)$$

and

$$y \geq 2\ ? \qquad (7.6)$$

Notice that the inequalities 7.1 through 7.6 are all conditions that have to be satisfied in one way or another. They provide a series of *bounds* that constrain acceptable combinations of answers for x and y. They do not tell you what x and y are, but rather what they cannot be.

How then do you find x and y? Can you try to "solve" these equations?

No! For a start, they are inequalities, not equations, and there are six inequalities for only two unknown variables x and y.

What about the objective of this exercise? Have you forgotten it? Can you still use it?

Remember, you were trying to make your total grade as high as possible.

Write the total grade you expect to achieve in terms of x and y.

How about

$$\text{Total grade} = 40 + \frac{10x}{3} + 40 + 5y$$

$$= 80 + \frac{10x}{3} + 5y \qquad (7.7)$$

Note that now you have a mathematical description of your problem: You want to maximize the expression in Equation 7.7 while satisfying all the conditions specified in 7.1 through 7.6.

This is probably unlike any problem you have encountered in your algebra classes. That should not worry you; the challenge is to find a way of getting to the answer!

See if you can develop your own algorithm for solving this mathematical problem. Think about the problem for 10 or 15 minutes and then read the questions below to see if they echo your thoughts or stimulate new ideas.

✓

Can you distinguish between values for x and y that are candidates for the solution and those values that are definitely not acceptable?

How can you ensure that your answer really does maximize the total grade?

What properties would you expect the answer to have?

Are there a series of steps that will lead you to the answer?

How could you draw a graphical representation of the problem? What axes would you choose? What could you plot?

Now go ahead and see if you can develop that algorithm!

✔

A Graphical Solution

A solution to this problem consists of a pair of values: x (the number of hours spent in the language laboratory) and y (time spent with the tutor). This suggests that it might be useful to represent what we know on a piece of graph paper, where one axis represents x and the other represents y. Since both x and y must be positive, we need to look only at the positive quadrant of the xy-plane, as in Figure 7.1.

P represents *any* point in that xy-plane. For example, P might be point (7,6). We want to know whether P is a *potential* answer to our problem. Is it?

When we talk about a "potential answer," we mean one that satisfies all the conditions 7.1 through 7.6. How could we represent those conditions on the graph paper?

✓

Take condition 7.1 for example. It specifies that $x + y$ must be less than or equal to 11. Suppose we draw the straight line

$$x + y = 11$$

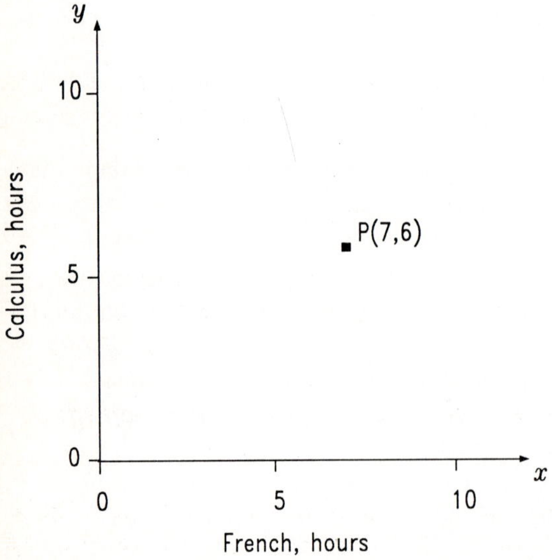

FIGURE 7.1
With this choice of axes, the point P represents 7 hours in the language laboratory and 6 hours with the Calculus tutor.

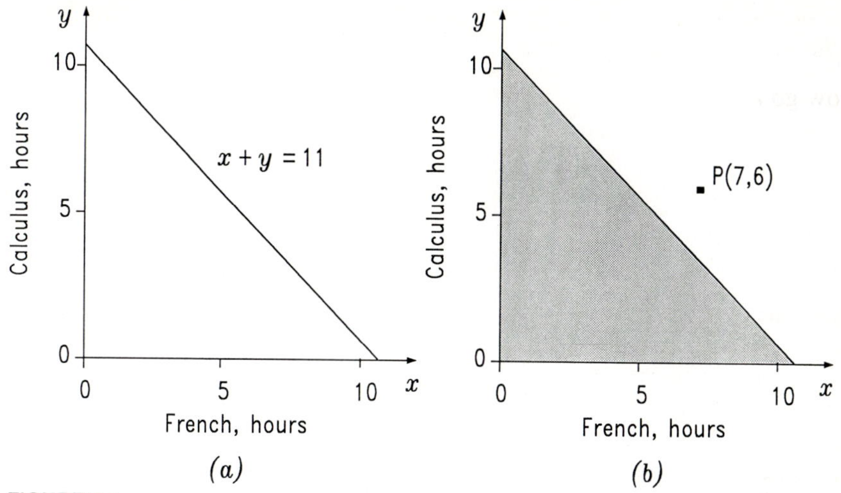

FIGURE 7.2
A graphical representation of the constraint imposed by your social calendar (Inequality 7.1). It reduces the set of acceptable solutions to the shaded area in part (*b*). It follows that P can immediately be excluded as an acceptable solution point.

as in Figure 7.2*a*. Then all points *below* that line will satisfy condition 7.1. It follows that we have *bounded* our solution: All potential answers to our problem must lie in the shaded triangle in Figure 7.2*b*. The point (7,6), for example, is not a potential answer.

Now look at condition 7.2, which specifies that x must be less than or equal to 8. In Figure 7.3*a* we have superimposed the straight line $x = 8$ on Figure 7.2*a*. Condition 7.2 tells us that only points to the *left* of that line are acceptable, and so our shaded area of potential answers is reduced to the quadrilateral shown in Figure 7.3*b*.

If we go on and add the rest of the conditions in this way, we eventually end up with Figure 7.4 and a small shaded polygon that contains all potential answers to this problem.

Does this approach remind you of the bounding heuristic?

It is in fact a very effective formalized application of that heuristic. We call the shaded area in Figure 7.4 the *solution space,* and any point in that area is a candidate for the solution.

"What happens," you may ask, "if there is no shaded area left after all the conditions have been applied?"

This is a good question because it is quite possible that a problem could be specified with conditions that rule out any answer at all. In that case we say

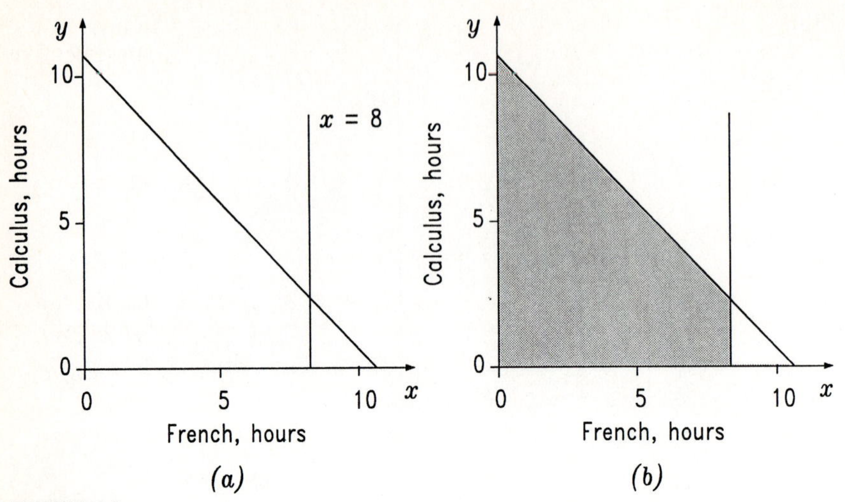

FIGURE 7.3
Here we add the constraint (Inequality 7.2) on the number of hours in the language laboratory (imposed by your limited attention span). Note that the shaded area has been further reduced.

that the problem is *infeasible,* or has no solution. A representation such as Figure 7.4 quickly demonstrates whether or not a problem of this kind has a solution. If it is infeasible, then one would have to relax one or more of the conditions. (Notice how this graphical representation would help you to argue about which condition or conditions to relax!)

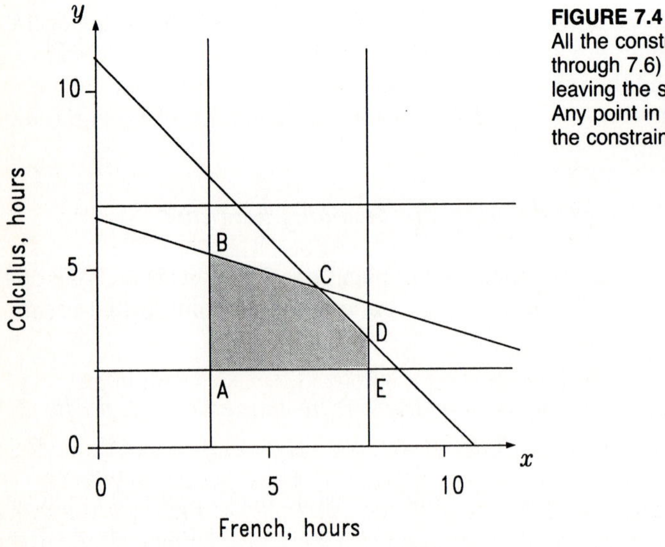

FIGURE 7.4
All the constraints (Inequalities 7.1 through 7.6) have been included, leaving the shaded polygon ABCDE. Any point in this polygon will satisfy the constraints.

Figure 7.4 has narrowed down our options, but of all the points in the shaded area, which one do we choose? Which *one* point satisfies our objective that it makes the expression

$$80 + \frac{10x}{3} + 5y$$

from Equation 7.7 as large as possible?

In the previous section we asked you to think about the properties of the solution to this problem. What conclusions did you reach? You should be able to reduce the solution space to just a few specific points. Have you any ideas on how to do this?

✓

If there were no conditions at all, we could make the expression in Equation 7.7 as large as we liked. At least one of the conditions must therefore set a limit to what we can do. We do not know which condition does that, but it follows that our solution must be a point on the boundary of the shaded area rather than a point inside the shaded area.

If we carry this argument one step further, once one condition has limited our answer, what prevents us from sliding up or down that particular bounding line until we make 7.7 as large as we like?

The answer must be *another condition*. Again, we do not know which, but it follows that our solution is not just any point on the boundary of the shaded area, it has to be a point where two bounding lines meet. In other words, it is a *corner* rather than an edge of the shaded area.

Figure 7.4 shows that there are only five corners, marked A, B, C, D, and E. Our answer to the problem must be represented by one of these points.

Which one? How are you going to find it?

You could calculate the coordinates of each point (remember, each point is at the intersection of two straight lines) and then substitute those coordinates in equation 7.7. Whichever gives the largest number is your answer.

Alternatively, you could identify the solution point graphically as follows:

1. Let z be equal to the expression we are trying to maximize, i.e., $z = 80 + (10x/3) + 5y$. (We call this the *objective function.*)

2. Choose any value for z, and plot the straight line corresponding to that value. Line R′R in Figure 7.5 is an example.

3. Choose another value for z, and plot the corresponding straight line. S′S in Figure 7.5 is an example. Notice that R′R and S′S are parallel. (Why?)

4. Take a straightedge, keep it parallel to R′R and S′S, and move it up until it is just about to move totally out of the shaded area. This happens when the

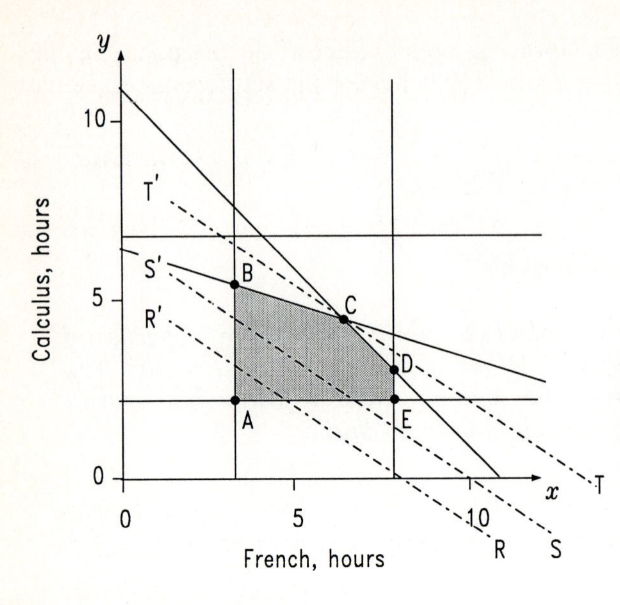

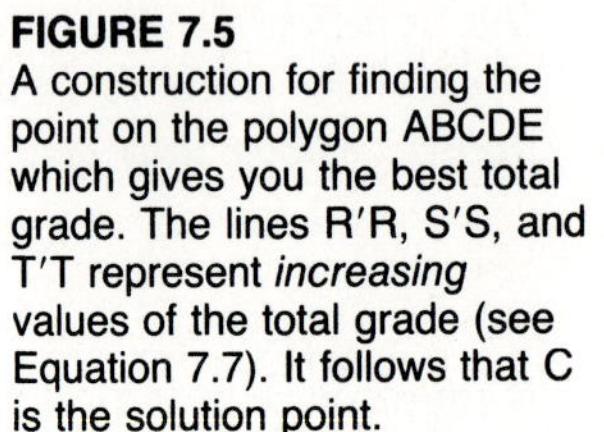

FIGURE 7.5
A construction for finding the point on the polygon ABCDE which gives you the best total grade. The lines R′R, S′S, and T′T represent *increasing* values of the total grade (see Equation 7.7). It follows that C is the solution point.

straightedge is in position T′T, just touching the point C. It follows that C must represent the solution to our problem.

Notice that C lies at the intersection of the time constraint 7.1 and the budget constraint 7.4.

Notice how this ties in with the very last argument we developed in the section entitled Solution by Argument.

We can find the x and y coordinates of C by solving the pair of equations (yes, "equations" this time because C lies at the intersection of two straight lines)

$$x + y = 11$$

and

$$5x + 15y = 100$$

WHAT IF?

So far we have been concentrating on how to get a solution. Now let us look at the solution itself. It tells you to spend 6.5 hours in the language laboratory and another 4.5 hours with the Calculus tutor. It promises you grades of 61.7 for French and 62.5 for Calculus, making a total of 124.2.

Suppose your friend offers to lend you $10. What difference would the loan make to your decision? How much better would your final grade be?

This is an example of a "what-if" type of question. What-if questions explore how your solution changes as you alter conditions or change assumptions.

Another example is: If you were to invest another hour of your time, how would you use it, and to what benefit?
Yet another is: Suppose the passing grade were increased from 50 to 70. How would this affect your decision?

What-if questions could even introduce new constraints.

What will you do if the tutor is only able to schedule 3 hours with you?

What-if questions can also change the objective.

How would you change your solution if the tutor were less effective (perhaps your Calculus grade would improve by only 4 points for each hour spent with the tutor)?

We leave you to explore these questions yourself.

The purpose in asking them is to show you how useful and versatile a representation such as that in Figure 7.5 can be. If you developed a different representation, how easily can you use it to answer these sorts of questions?

✔

DISCUSSION

What Is "Different" about This Problem?

Notice that although there is a lot of information in the description of this problem, the actual answer is determined by only two conditions and the objective function.

If we knew which were the crucial conditions beforehand, this would be an easy problem to solve. What makes it difficult (and interesting) is that we do not know which conditions are going to be crucial. Moreover, if we slightly change the problem (or ask what-if questions) we could easily find that a different two conditions become crucial.

If you happened to guess which were the two important constraints, read the previous paragraph again. You were lucky, and relying on luck is not a good heuristic.

It follows that our solution method is a kind of search, one that aims to identify the crucial conditions.

Which Is the Preferred Method of Solution?

Which method did you use? Which do you prefer?

How easy would it be to use your preferred method if we made the problem more complicated? Suppose we added another three or four conditions? Suppose we added a third alternative (such as working through Calculus exercises with a friend)?

Having recognized that the solution to this problem is a kind of search, it follows that a preferred method of solution is a search procedure that leads to the answer in an organized, reliable, and efficient manner. Table 7.1 will not do that for any but the most simple problems of this kind. Solution by argument will also not be a good method; one could so easily end up arguing in circles or losing track of the logic. So we prefer to express the problem mathematically and solve it graphically.

What are the key features of this solution?

1. First we identify our variables.

2. Then we express the conditions in terms of these variables, as in conditions 7.1 through 7.6. We call these the *constraints.*

3. We graph the upper or lower limits of each constraint, as in Figure 7.4, and determine the *solution space.*

4. Next, we express our objective in terms of an *objective function* that we wish to *maximize* (or, perhaps in a different problem, *minimize*). We find where the objective function just touches the boundary of the solution space to determine our solution point.

What would you do if you were trying to minimize the objective function? (Notice that the minimum solution to this problem is at point A in Figure 7.5, where you only just manage to scrape through both quizzes!)

5. Finally, it is often useful to perform what is known as a *sensitivity analysis,* i.e., to ask a series of relevant what-if questions.

"But what do you do," we can hear you ask, "if you have three or more variables?"

The graphical solution breaks down, but the ideas behind it can be incorporated into a systematic search from corner to corner of the three-, four-, or *n*-dimensional solution space. The *simplex algorithm* (see the end of this chapter for a reference) is one that performs this search reliably and efficiently and can easily be programmed for a computer.

The crucial step is still: Express the problem in mathematical terms.

What assumptions are built into this approach?

Notice that all the constraints and the objective function can be represented in terms of straight lines in the case of two variables (as in Figure 7.5). The approach would not necessarily work if this were not the case. That is why the technique we have developed here is called *linear programming.*

Learning to Use Standard Techniques

Linear programming is a standard technique or tool in the problem solver's toolkit. We hope that you have developed an appreciation, from this exercise, of how the technique works and where it works, i.e., the type of problem that can be solved by this technique.

Notice how we have exposed you to linear programming. A regular textbook would probably have started out by explaining the linear programming algorithm and would have followed up by giving you exercises to do. We deliberately took a different approach. As in other chapters of this book, we preferred to create an environment where you could conceivably develop linear programming for yourself. If you did not discover it, then you could watch us as we exposed it and could compare it with whatever approach you developed for this problem. Either way, we expect you to come away with a better understanding of why linear programming is a useful technique.

Once you have added this technique to your problem solving toolkit, it adds to the power of the heuristic: Is there a standard technique that I can use to solve this problem? When you are confronted with a fresh problem, you can ask yourself whether it is useful to try to think of the problem in linear programming terms. Is this a problem where you are trying to maximize or minimize a function of two or more variables? Are there conditions that can be described as constraints? Can you write these conditions in terms of linear relationships?

Linear programming, like any other tool, can be a double-edged sword. If you have a problem that can be specified as a linear programming problem, then you have ready-made programs and theorems at your disposal. It would be sad if you did not know about linear programming and tried to tackle it in a less efficient way. On the other hand, there is the danger of "the hammer syndrome" (if the only tool you have is a hammer, then every problem gets treated like a nail): you could be so anxious to recast a problem as a linear programming problem that you distort it out of all recognition, and your solution, however efficient, does not really address the original reason for building a model.

There are algorithms available for solving problems with nonlinear constraints and/or objective functions. However, since linear programming techniques are more powerful than nonlinear algorithms, there is much to be gained by simplifying constraints that are nonlinear: it is often relatively easy to replace a curve by a more-or-less similar straight line. Here again, the simplification is useful provided it does not distort.

If you think about it, we made some important simplifications when we wrote down our objective function. How?

✓

We argued, for example, that you would add 3⅓ points to your French score for every hour spent in the language laboratory. There must be a limit to this, otherwise you could end up scoring more than 100 percent. In practice, as you spend more and more time in the language laboratory, you will eventually hit diminishing returns. Points scored versus time in the laboratory is therefore likely to be a curve, like that shown in Figure 7.6, rather than a straight line.

We have assumed that you are unlikely to spend enough time in the laboratory to notice the diminishing return. Our approximation to the curve is the straight line in Figure 7.6. It would be prudent to check that our answer is not on the dotted (unrealistic) section of the straight line!

Why Not a Stochastic Model?

Did you wonder at any stage whether to build a deterministic or stochastic model? Where does uncertainty come into this problem? Are there ways of dealing with it?

You cannot be sure that an hour spent in the language laboratory will really improve your French grade by 3⅓ percentage points. Nor can you be sure that

FIGURE 7.6
The curve is a more realistic relationship between your expected French grade and the time spent in the language laboratory than the straight line used in this model. The dashed portion of the straight line is therefore unrealistic.

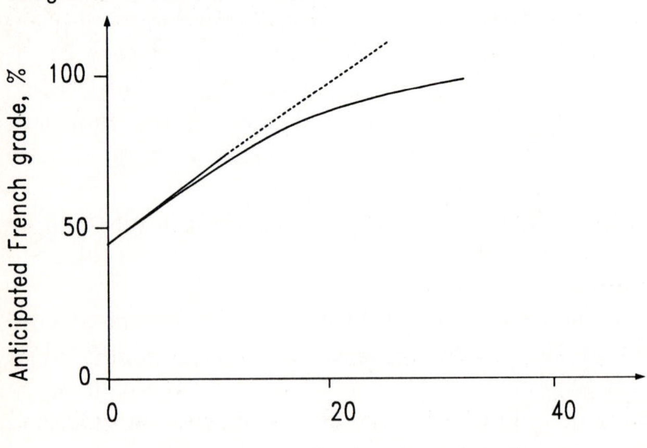

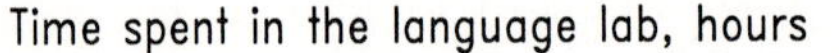

you would score 40 percent without any further preparation for the French quiz. There is a similar uncertainty about the outcome (remember we talked about the "likely" outcome) of the Calculus quiz.

But this uncertainty is not a crucial part of the problem. We would confuse rather than improve our model if we tried to make it stochastic. The numbers we are using ($3\frac{1}{3}$ points per hour, 40 percent, and so on) are guesses or estimates rather than quantities that should be represented by probability distributions. We can deal with the uncertainty in these numbers in other ways.

For example, we could ask: "What would happen if our estimate of 40 percent was too high?"

How sensitive is our solution to that number?

The number appears in two places:

1. Explicitly, in the objective function 7.7.
2. Implicitly, in the constraint 7.5. This constraint should really be written

$$\frac{10x}{3} \geq 50 - 40$$

Suppose we replace 40 by v in both places. Then a useful question to ask is: "How small can we make v without changing our decision point?"

Will the introduction of v in the objective function make a difference to the solution point? Will it ever make us choose a point other than C in Figure 7.5?

No, because it does not affect the *slope* of the lines R′R, S′S, and T′T.

But it does make a difference to the constraint. The equation for the line AB becomes

$$\frac{10x}{3} = 50 - v$$

What happens to the line AB in Figure 7.5 as we decrease v?

It moves further and further to the *right.*

How far could we move it without changing the solution point?

Until AB passes through the point C. As soon as it moves to the right of C, the solution will change.

What is the x coordinate of C? What is the value of v when AB passes through C?

The x coordinate of C is 6.5. If AB were to pass through C, its equation would therefore be

$$10\frac{6.5}{3} = 50 - v$$

or

$$v = 28\tfrac{1}{3}$$

In other words, we could change our estimate from 40 to 29 before it would affect our solution!

What about your estimate of 40 percent for the Calculus quiz?

Arguments such as this are far more useful than a stochastic model could ever be for this type of problem.

FURTHER READING AND SIMILAR PROBLEMS

In this chapter we have seen the graphical solution to linear programming problems and have mentioned the simplex algorithm for solving similar problems with more than two variables. *An Introduction to Operations Research Techniques* by H. G. Daellenbach, J. A. George, and D. C. McNickle (Boston, Allyn and Bacon, 1978) has a section on linear programming that begins with the graphical approach and then goes on to develop the simplex algorithm.

Look back at the scheduling problem of the previous chapter. Can you formulate the problem as a linear programming problem? What are your variables? What are your constraints? What are you trying to maximize (or minimize)?

The book by Daellenbach, George, and McNickle describes how to formulate network flow problems as linear programming problems. It also discusses the solution of transportation problems such as the following:

A Transportation Problem

Sandbaggers, Inc., is a company that purchases silica sand, cleans it, and then sells the silicon dioxide to manufacturers of computer chips. Sandbaggers, Inc., operates two cleaning plants, called A and B. They obtain the silica sand from three independent suppliers (Sam, Alan, and Steve) who are willing to supply the sand in the following amounts and at the following prices:

Sam	200 tons at $10 per ton
Alan	300 tons at $9 per ton
Steve	400 tons at $8 per ton

Shipping costs in dollars per ton are as follows:

	To	
From	**Plant A**	**Plant B**
Alan	2	2.5
Sam	1	1.5
Steve	5	3

Plant capacities and labor costs are as follows:

	Plant A	**Plant B**
Capacity	450 tons	550 tons
Labor cost	$25/ton	$20/ton

The cleaned and processed silicon dioxide is sold at $50 per ton to the chip manufacturers. Sandbaggers, Inc., can sell at this price all they can produce. How should the company plan its operation at the two plants so as to maximize its profit?

An Economic Problem

A building contractor has acquired a 10-acre site and plans to develop it as a residential area. The contractor has two basic designs for homes: a low-income home and a middle-income home. He expects to take 5 years to complete the project.

Low-income homes cost $30,000 to build at a density of 20 homes per acre, whereas middle-income homes cost $40,000 to build at a density of 12 homes per acre. The contractor expects to sell a maximum of 100 low-income homes and 75 middle-income homes, but the zoning regulations will allow no more than a total of 155 homes on the site. The local city council has also stipulated a formula to ensure that sufficient low-income homes are constructed. This formula stipulates: The number of low-income units shall be at least 35 units greater than half the number of middle-income units.

The contractor employs a work force of 60 people. The architect estimates that it will take 2.0 labor-years to build a low-income home and 2.5 labor-years to build a medium-income home.

The contractor expects to sell a low-income home for $50,000 and a medium-income home for $70,000.

Your task is to advise the contractor. How many of each type of home should he build?

Linear Programming and Wildlife Management

For a somewhat different application you may want to look at the chapter entitled, Cropping Strategies and Linear Programming, in *Building Models for Conservation and Wildlife Management,* by Anthony Starfield and Andrew Bleloch (published by Macmillan in New York in 1986 but now marketed by McGraw-Hill).

Linear programming emerges as a tool for looking at how to control a herbivore population in a game preserve. The example highlights both the advantages and disadvantages of a linear programming model, and the authors develop the case for several alternative representations.

CHAPTER **8**

A CAB CONTROL SYSTEM

BACKGROUND

S & B is a successful and progressive engineering design company. It was recently awarded a contract for the design of an automated monorail system for a city. The concept is one that envisages small, computer-controlled cabs moving at high speeds on a network of rails.

There were many aspects to this project (cab design, power transfer system design, cab control system design, sensor design, and so on), and S & B organized teams to address each area.

The Software and Control team was assigned the task of developing and testing computer software for the cabs. Each cab, they were told, would be autonomous and controlled by on-board software. Their job was to design on-board routines to drive the cabs safely and efficiently, even on congested rails.

The leader of the team first checked with engineers in the sensor design group and was satisfied that reliable sensors would be available for monitoring speed and distances to or from specific points on the track or between one cab and another.

The team leader then drew up a plan for developing a whole series of controlling routines.

Take a few minutes to think about a journey in one of the cabs. Break the journey down into situations that might require different controlling routines. Make a list of those situations.

✓

Figure 8.1 is from a document drafted by S & B. It identifies five different routines and shows schematically where each would apply. The five routines include:

1. A routine for starting a cab and getting it up to speed (on a special loading sideline).
2. A routine for merging traffic at a junction.
3. An "open line" routine for a cab that can detect no other cabs on the line ahead.
4. A "following" routine for a cab that can detect at least one other cab on the line ahead of it.
5. A routine for exiting onto a loading sideline and stopping.

What should these routines actually control? What will they adjust? Think about the controls that you use when you drive a car. What, for instance, do the pedals on your car adjust?

FIGURE 8.1
From the files of S & B. This sketch depicts different traffic situations that the automated monorail system will have to resolve and control.

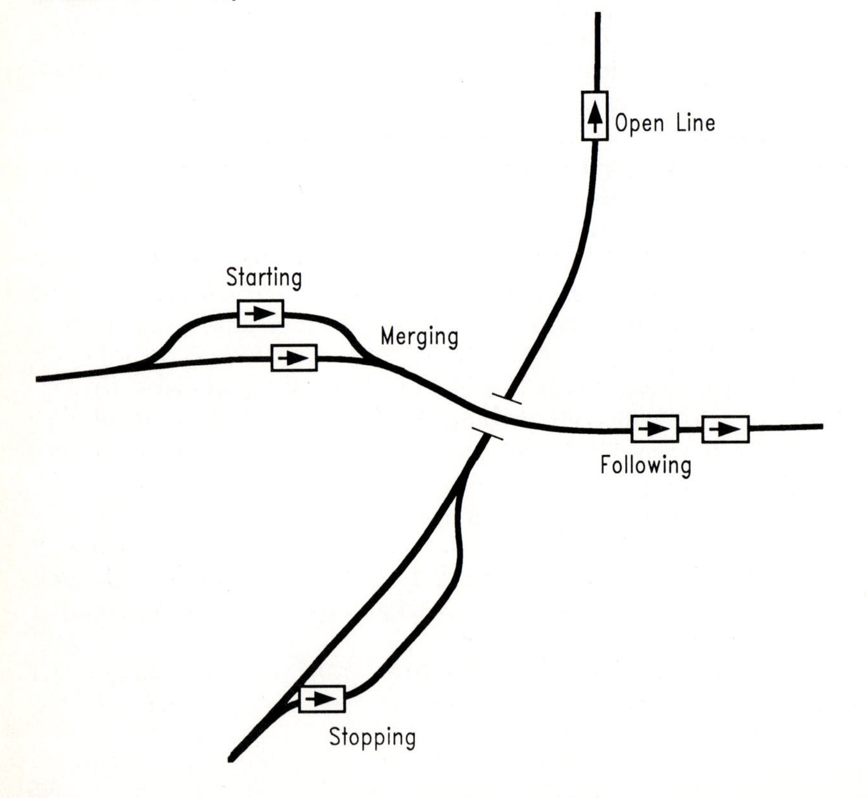

Simplistically, when you depress the gas pedal, you accelerate the car; lift your foot off the gas pedal, or depress the brake pedal, and you decelerate the car.

The design engineers at S & B decided that their computer routines would adjust the acceleration or deceleration of a cab.

The "Following" Routine

The Software and Control team ran into difficulties with the "following" routine for two or more cabs on the same line.

They agreed on some design features:

- Each cab would have a sensor that searched for other cabs within a certain distance on the line ahead.
- When this sensor detected a cab ahead the controlling routine would switch from "open line" to "following" mode.
- The "following" mode algorithm would accelerate or decelerate a cab depending on its relationship only with the cab *immediately* in front of it. Cabs would not be able to detect what was happening behind them.

Think again about driving a car. What algorithms do you use to avoid colliding with the car in front of you? Take 1 or 2 minutes to describe the strategy that you use.

✓

Do you monitor the distance between yourself and the car ahead? Or do you try to judge whether you are driving faster (or slower) than the car in front of you?

The Software and Control team was split down the middle over which algorithm to use for accelerating or decelerating a cab in "following" mode. At this time there are two opposing factions in the team.

One group believes that it is only necessary to monitor the *distance* to the cab ahead. When this distance is greater than a preset distance D, they want to accelerate the cab. Conversely, they will decelerate it if the distance is less than D. They call this the *safe-distance* design and argue that it is possible to choose D with a built-in safety factor that prevents collisions.

The other group argues that it is essential to monitor the *speed of approach* (S). Their algorithm will decelerate the cab whenever S is positive (the two cabs are moving closer together) and accelerate it when S is negative (the two cabs are moving further apart).

Neither group is willing to concede that there are merits to the other group's approach. Discussions have become emotional rather than rational, and the Software and Control project is behind schedule.

YOUR PROBLEM

S & B has given up on resolving this problem internally and has hired you as a qualified and impartial modeler to investigate the merits and demerits of the proposed algorithms.

At your first meeting with S & B it emerges that the two groups have been so busy fighting that neither has actually worked out detailed proposals. They want you to begin by developing two alternative algorithms for controlling a cab in "following" mode. The first must be based on the ideas of the safe-distance group, while the second must incorporate the ideas of the speed-of-approach group.

S & B have presented you with a major assignment. Developing the two algorithms is only the first of a number of tasks. Before you start working on it, it would be a good idea to make sure that you fully understand the "big picture."

Take 10 to 15 minutes to consider what you will do with the algorithms after you have developed them. How are you going to assess their advantages and disadvantages? Make a list of the various tasks you expect to perform. List them in order so that you can be sure that you anticipate how one task will build on another.

✔

One possible breakdown of the problem is to:

1. Develop control strategies or algorithms that are representative of the two opinions.

2. Develop a "computer testbed" for the algorithms. An example of a testbed for, say, a car engine would be a corner of the factory where the engine could be mounted with all the associated fuel pipes, electrical connections, and test equipment. Each engine would be mounted in turn and its performance measured as if it were actually in a car. It would really be useful to have a computer program to model the environment of the cab in an analogous way. Different algorithms for controlling the cab could then be inserted in this program and their performance recorded and compared.

3. Decide what tests you will want to make in your computer testbed. In other words, plan a series of tests that will allow you to evaluate the relative performance of the different control algorithms.

4. Execute the tests and draw conclusions from them.

5. Compile information from tasks 1 through 4 to support your conclusions, and prepare a report for S & B.

Does your list differ substantially from ours?

In practice, you would have drawn up your list as a way of confirming with S & B that what you were planning to do was what they wanted you to do. We have drawn up ours as a way of telling you what we expect you to do with this problem.

In both cases, writing down an ordered list is an efficient form of communication. It is also a useful habit to develop whenever you have a complex task. It helps you to identify what needs to be done and to keep each step in perspective. It also helps you to allocate time and resources; you do not want to do a fantastic job on task 1, for instance, and then run out of time before you get to tasks 4 and 5. This may seem obvious, but it's surprising how often it's overlooked.

YOUR FIRST TASK

Now you are ready to tackle your first task. Remember, you were asked to develop two algorithms for controlling a cab, one based on the concept of safe distance, the other on speed of approach.

During your next meeting at S & B you are expected to describe the two algorithms. You will also have an opportunity to ask the engineers for further information.

Your ideas for the algorithms may be preliminary at this stage, but they should be complete. You should prepare, in point or flowchart form, unambiguous instructions for controlling a cab in "following" mode. You should also prepare a list of questions you want to ask the engineers.

We suggest that you spend somewhere between 30 and 40 minutes on this task. In the next section we have compiled a list of points you might have considered. Look at it only after you have finished or if you need assistance.

✔

POINTS YOU MIGHT HAVE CONSIDERED

- *How many cabs did you decide to include in your algorithm? Why?*
- *How frequently should the sensors monitor distance and/or speed of approach?*
- *Did you draw a diagram that shows what information goes into your control algorithm and what information comes out of it?*
- *What are the essential variables? What thought have you given to choosing a notation? Why do you like your notation?*
- *What is the maximum acceleration of a cab? What is the maximum deceleration?*
- *Have you kept your algorithms simple and easy to implement?*

If you are still happy with what you have done after reading this, go ahead and read how we developed our algorithms. Otherwise spend more time preparing yours.

✔

OUR CONTROL ALGORITHMS

General Considerations

We decided that we only needed to consider two cabs at this stage: the cab we are controlling (call it cab B) and the cab ahead of it (cab A). Thinking ahead to our next task, we will eventually want to "drive" cab A ourselves in our computer testbed, while different algorithms are used to control cab B. We will want to monitor how cab B responds to the way in which we drive cab A.

Having decided to consider only two cabs, we tried to draw a diagram that would represent any control algorithm for cab B. Figure 8.2 is the result. It shows the algorithm as a box (we still have to decide on the details) with inputs and outputs. In general, the inputs are the positions and speeds of the two cabs. The output is the acceleration or deceleration (which is just negative acceleration) for cab B.

How often should we adjust the acceleration of cab B? Are we going to check and adjust at regular time intervals, or only when situations arise that require action? In other words, do we want a time-driven or event-driven algorithm?

One way to answer these questions is to think of driving a car. A good driver checks and adjusts all the time, not only when there is a potential crisis. This suggests that we should choose a fixed time interval; let us call it *dt*. Again, if we think of driving a car, especially at high speeds, we will want *dt* to be small: perhaps a fraction of a second or, at most, one or two seconds.

The first questions on our list for the engineers at S & B are "How fast are your sensors?" and "How small can we make *dt*?"

How will we use Figure 8.2 over and over again? What will we know? What will we have to calculate?

Suppose we are at the beginning of a time interval *dt*. Suppose we know the positions and speeds of both cabs at that time. Our control algorithm will then

FIGURE 8.2
A general representation of the inputs and outputs for the algorithm controlling cab B as it follows cab A.

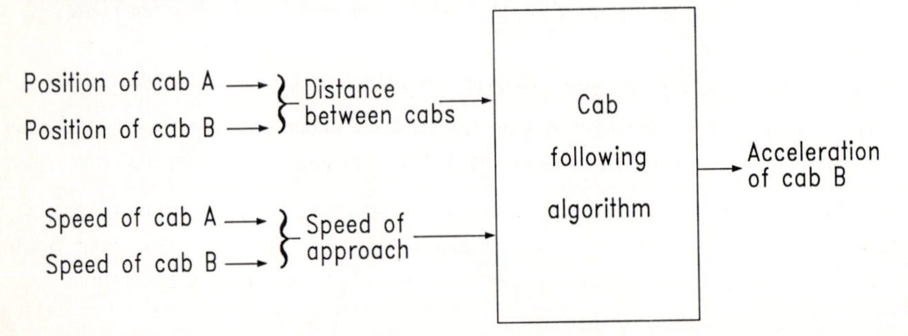

calculate and adjust the acceleration (or deceleration) of cab B. At the end of the time interval (or the beginning of the next time interval, depending on how you look at it), we will need to know the positions and speeds of both cabs again.

Have we identified all our variables? What notation does the above discussion suggest?

Suppose we start modeling the movements of the two cabs at a time which we call $t = 0$. A subscripted notation would be a convenient way of describing the positions and speeds of the two cabs at times dt, $2dt$, $3dt$, and so on. In general, after n time intervals (at time ndt) we could let

x_n = position of cab A (we can measure its position along the track from some arbitrary starting point)
y_n = position of cab B (measured from the same arbitrary starting point on the track)
u_n = speed of cab A
v_n = speed of cab B

These would be the inputs for our control algorithm at the beginning of the *next* time interval (the one starting at time ndt). Our algorithm would instantaneously calculate and adjust the acceleration of cab B. We could call this acceleration a_n. If the response mechanisms are sufficiently fast, a_n will be the acceleration applied to the cab throughout the time interval starting at ndt and ending at $(n + 1)dt$. (We need to ask the engineers about response times!)

Notice how introducing a notation clarifies what we are trying to do. Our current task is to describe how the control algorithm should calculate a_n, given x_n, y_n, u_n, and v_n.

We begin by considering the safe-distance approach.

Safe-Distance Control Algorithms

In terms of the notation we have introduced, the distance between the two cabs after n time intervals will be $x_n - y_n$. (Why not $y_n - x_n$?) If this is less than the safe distance D, then we want to decelerate cab B, otherwise we want to accelerate it. There are a number of different ways we could do this. For example, we might have an algorithm that says

1. If $D > x_n - y_n$, then apply maximum acceleration
2. If $D = x_n - y_n$, then leave the speed unchanged
3. If $D < x_n - y_n$, then apply maximum deceleration

Suppose a_{max} is the maximum acceleration and d_{max} is the maximum deceleration. (We must remember to ask the engineers at S & B to give us actual values for both a_{max} and d_{max}.) The above algorithm can then be written:

Algorithm 1

1. If D > $x_n - y_n$, then $a_n = a_{max}$
2. If D = $x_n - y_n$, then $a_n = 0$
3. If D < $x_n - y_n$, then $a_n = -d_{max}$

What algorithm did you devise? Was it similar to this? Do you think this is a good representation of the "safe-distance" philosophy? Are we doing justice to their case?

We think not! This algorithm might be worth testing, but it has one very serious weakness. What do you think it is?

✓

Suppose that cab B is slightly too close to cab A. The above algorithm will apply maximum deceleration, with the result that at the next time step cab B is likely to be slightly too far behind cab A, and so the algorithm will apply maximum acceleration.

This could lead to a very uncomfortable ride. Do you know anybody who actually drives like this?

How do we remedy this?

Well, if B is just a little too close to A, we only want to decelerate slightly. If you have not already done so, write down an algorithm that would accomplish this.

✓

A finer control is one where the acceleration is made proportional to the difference between the actual distance between the two cabs ($x_n - y_n$) and the desired distance (D). A formula that does this is

$$a_n \text{ is proportional to } (x_n - y_n) - D$$

or

$$a_n = k[(x_n - y_n) - D] \tag{8.1}$$

where k is a proportionality constant.

Why only one formula when we previously had three different cases?

Notice that Equation 8.1 automatically takes care of all three cases. Notice too that we can think of k as a design parameter; we can adjust its value to improve the design in some desired way.

Is this it, then? Is Equation 8.1 a good algorithm for the safe-distance group? If not, can you improve it?

There is still a hitch. If the cabs are too close (or too far), it is possible that Equation 8.1 might tell the cab to decelerate (or accelerate) faster than it can. A better version of this algorithm is

Algorithm 2

1. Let $a_n = k[(x_n - y_n) - D]$
2. If $a_n > a_{\max}$, then reset a_n to $a_{\max}$
3. If $a_n < -d_{\max}$, then reset a_n to $-d_{\max}$

We have presented two possible algorithms for the safe-distance group. There are of course others. If yours is different from either of the above two, the chances are that it is more complicated. If so, do you have good arguments in favor of the extra complication?

A Speed-of-Approach Control Algorithm

In terms of our notation, the speed of approach (S) of cab B at time ndt will be equal to $v_n - u_n$. (Are you sure it is not $u_n - v_n$?)

Can you write down an algorithm that is similar to Algorithm 2 but depends only on the speed of approach S?

Our speed-of-approach algorithm is:

Algorithm 3

1. Let $a_n = K(u_n - v_n)$, where K is a design parameter
2. If $a_n > a_{\max}$, then reset a_n to a_{max}
3. If $a_n < -d_{\max}$, then reset a_n to $-d_{\max}$

Did you write out your algorithms so they were obvious and unambiguous, as we hope ours are?

"But," we can hear you say, "My algorithms are more complicated than yours."

All the more reason to write them out clearly! If your algorithms are indeed more complicated than ours, we recommend that you ask yourself whether they are too complicated. An even better question to ask yourself is "What is the advantage of the added complication?"

YOUR NEXT TASK: BUILDING A COMPUTER TESTBED

The engineers at S & B think your control algorithms are worth testing. Your next task is to build the computer testbed for testing and comparing different algorithms.

Begin by listing the specifications for the testbed. What must the computer program do?

Our specifications are:

1. We will provide the initial positions and speeds of the two cabs, the speed (at all times) of cab A, and of course the control algorithm.

2. The program must calculate the acceleration of cab B at the beginning of each time step and then compute the following at the end of each time step:

- The position of cab A
- The position of cab B
- The speed of cab B

3. The program must be capable of tabulating or graphing the positions and speeds of the two cabs.

The "guts" of this program is in specification 2 above. If we can work out how to make the calculations for one time step, then we can make them for all time steps. Suppose you know the positions x_n and y_n and the speeds u_n and v_n. How can you calculate a_n and then x_{n+1}, y_{n+1}, and v_{n+1}? (Remember, you already know u_{n+1} from specification 1 above.)

✔

Calculating the acceleration a_n is easy: whichever algorithm we use should tell us exactly how to make the calculation.

Let us first look at cab B. The engineers at S & B have assured us that their measuring and control devices all operate and react in milliseconds rather than seconds. It follows that we know the position of cab B (y_n) and its speed (v_n) at the beginning of the time interval. We also know that its acceleration (a_n) is *constant* throughout the time interval dt. We want to calculate its position and speed at the end of the time interval.

Do you know any formulae that apply to bodies that move with constant acceleration? If not, can you argue how much speed the cab will gain if an acceleration a_n is applied for a time dt? Can you argue how far the cab will move in that time?

✓

Since the acceleration is constant throughout the interval dt, the speed of cab B at the beginning of the next time interval will be

$$v_{n+1} = v_n + dt\, a_n \tag{8.2}$$

Notice that if the control algorithm tells the cab to decelerate, a_n will be negative and so v_{n+1} will automatically be less than v_n.

The speed of cab B is *not* constant during the interval dt. However, if we were to plot its speed versus time during the interval dt, we would get a straight line (with slope equal to the acceleration). We can therefore calculate how far cab B has moved from the average speed during the time interval:

$$y_{n+1} = y_n + dt \times \text{average speed during the time interval}$$

$$= y_n + dt\frac{v_n + v_{n+1}}{2}$$

$$= y_n + dt\frac{v_n + v_n + dt\, a_n}{2} \qquad \text{(by Equation 8.2)}$$

therefore,

$$y_{n+1} = y_n + dt\, v_n + 0.5\, dt^2\, a_n \tag{8.3}$$

The argument we have just used for calculating the position of cab B can be repeated for cab A. Its position at the end of the time interval will be

$$x_{n+1} = x_n + dt \times \text{average speed during the time interval}$$

and since we know its speed both at the beginning and end of the time interval, we can write

$$x_{n+1} = x_n + dt\,\frac{u_n + u_{n+1}}{2} \tag{8.4}$$

Did you manage to derive Equations 8.2, 8.3, and 8.4? If not, did you recognize that these were the kinds of equations you would need? Did you look for somebody who could help you develop them?

The key is to be able to update the speed and positions of the two cabs from one time interval to the next. If you build these equations into your computer testbed, you will be able to simulate the movements of the two cabs for any given control algorithm.

Now go ahead and build your computer testbed! We do not care how you do this: you might prefer to write your own program or you could use a spreadsheet or commercially available modeling software. However, this is one exercise where you really have to go to the computer. The exciting part of this exercise is still to come: it is not in developing the control algorithms, but in *testing* and *comparing* them.

✔

PLANNING YOUR TESTS

Choose only two algorithms, one of the safe-distance type and the other of the speed-of-approach type. If you are still happy with algorithms that you have developed, use those. If you wish to modify your algorithms, go ahead and do so. Otherwise choose two of our algorithms.

Your computer testbed is waiting for you. You are going to test the two algorithms thoroughly and report to S & B on their performance.

As a first step, it would be prudent to plan the tests you intend to make. Draw up a list of points you should think about before you start testing the algorithms. Then compare your list with ours.

✓

1. What types of tests should you make? In other words, how will you drive cab A?

2. How will you choose the time interval *dt* and values for design parameters such as k and K?

3. What results should you print out or plot?

4. How will you evaluate the different strategies? What is your basis for saying that one algorithm is better than another?

5. So far we have concentrated on only two cabs. How many cabs should be included in the test runs? Do you need to modify your computer testbed to accommodate a line of traffic?

YOUR FINAL TASKS

Now that you have planned your tests, the time has come to

- Run them
- See what you can learn from them
- Compile your report for S & B

You have heard that model train enthusiasts at S & B are planning to build a miniature track and miniature versions of the cabs to test out the competing strategies. They are putting a great deal of effort into this; they have a large budget, and you suspect that they are having lots of fun. You view this as a challenge; it would be exciting if you could draw your own conclusions and submit your report long before theirs (and on a much smaller budget too). After all, your computer testbed is designed to produce the results they are looking for.

In fact a computer model has an important advantage over a miniature track and cabs. It is so much easier to modify and adjust. When you implement your tests, make sure that you take full advantage of the flexibility that a computer model provides. Look carefully at all the results. You may see trends which pro-

voke ideas and suggest modifications or additional tests you would like to make.

You could spend several hours using your model to evaluate and compare the two strategies. When you have done so, select the most important results, and use them to illustrate what you have learned and to support your recommendations to S & B.

✔

OUR RESULTS

Did you find your computer world as fascinating as the miniature monorail system the engineers wanted to build?

What did you learn from your model? Write a paragraph summarizing your main conclusions.

What were the advantages and disadvantages of each algorithm? Did you come out strongly in favor of one group or the other?

How convincing was your presentation to the engineering firm?

Our Approach

We built a computer testbed that could handle up to four cabs. It allowed us to drive the front cab and then applied whatever algorithm we chose to the remaining cabs. Each time we tested an algorithm, we first looked at only two cabs (to make sure that it was working properly) and then added the other cabs (to see how a line of traffic would behave).

The next decision was how to drive cab A. We chose three standard test patterns:

1. Drive cab A at a constant speed.

2. Start with cab A moving at a constant speed and then apply one-half the maximum deceleration until cab A has come to rest.

3. Repeat the first test but add a random component to the speed, the idea being to test the algorithm when the front cab is either malfunctioning or is itself adjusting speed in response to traffic ahead.

We sidestepped the choice of time interval dt and the design parameters. We decided we would tune these in once our program was working, i.e., we would change values and see what difference that made.

How to compare the algorithms? We decided to test the following features:

1. *Safety*. We wanted to check that cabs never collided or got dangerously close. We therefore printed out the distances between cabs at every time interval.

2. *Efficiency*. We decided to print out the speeds of all cabs to get an idea of how fast the traffic was moving.

3. *Comfort.* (Did you remember that there were people inside the cabs?) We printed out the accelerations of all cabs (to make sure that the maximum was never exceeded) and the *rate of change* of acceleration.

Why do you think we are interested in the rate of change of acceleration?

Our Results: (1) The Safe-Distance Algorithm

We spent a great deal of time and effort trying to "tune in" the parameters for the safe-distance algorithm. No matter what values we chose for the time interval dt, the parameter k, and the safe distance D, we could not get cab B to behave in a reasonable way!

Figure 8.3 shows the distance between the two cabs for a computer run. Our algorithm, remember, is designed to maintain that distance at a value close to D, which in this case was 30 m. The figure shows how dismally we failed. In fact the distance between the two cabs even becomes *negative.* (How does one interpret that? We eventually realized that cab B had crashed through cab A and was now ahead of it!)

Figure 8.4 is a plot of the speed of cab B. Notice that the speed becomes negative: cab B is actually reversing. We have seen some bad drivers, but never anything quite as bad as that!

Eventually we discovered that the safe-distance algorithm worked perfectly if we started with the two cabs moving at identical speeds at a distance precisely D apart. Any slight deviation in speed or distance, however, caused cab B to behave rather like a nervous driver. It never quite matched speed with

FIGURE 8.3
How the distance between two cabs changes with time when the safe-distance algorithm is applied. Remember, the algorithm is designed to maintain a 30-meter separation!

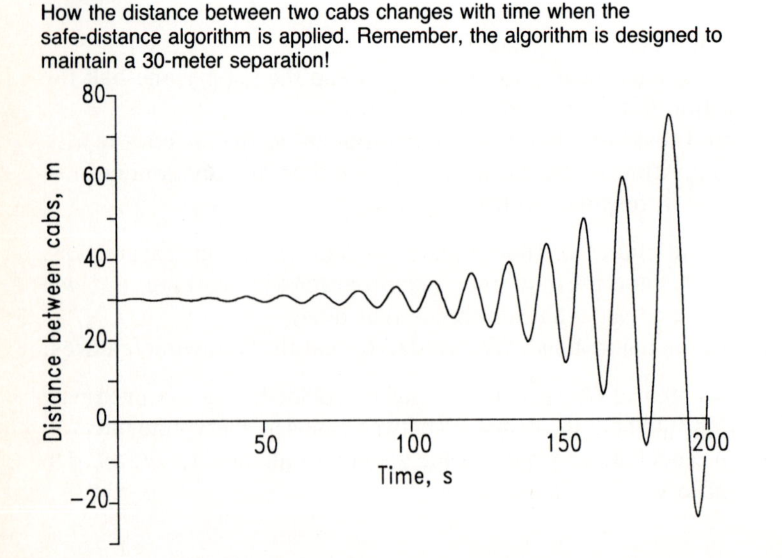

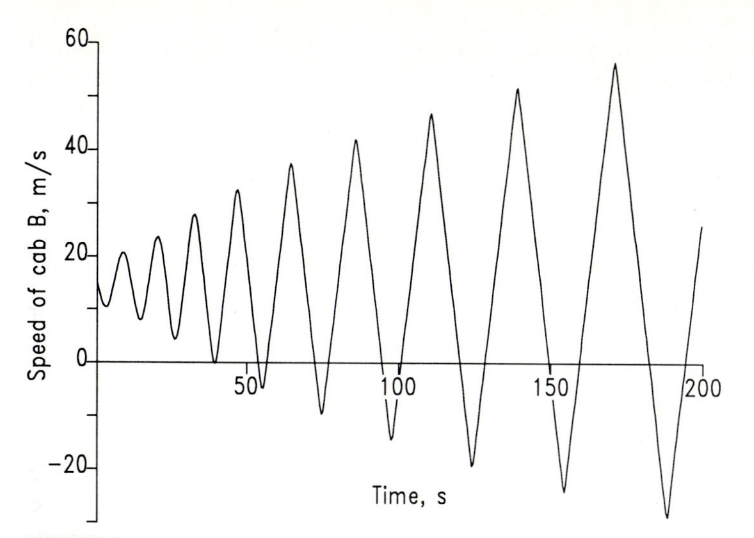

FIGURE 8.4
The speed of cab B as a function of time for the safe-distance algorithm. Cab A is moving at a constant speed of 15 m/s.

FIGURE 8.5
With a careful choice of parameters in the safe-distance algorithm, fluctuations (the wobble) in the speed of cab B can be reduced. Compare this with Figure 8.4.

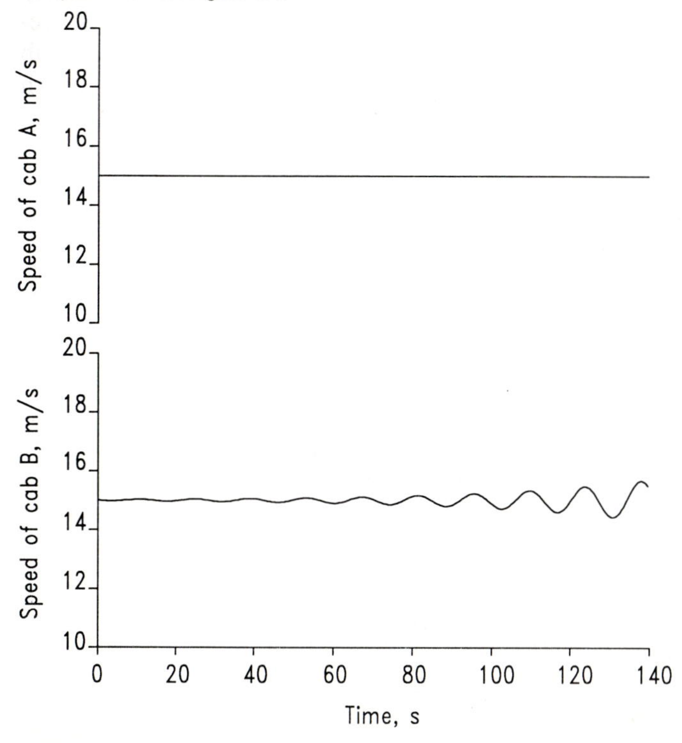

cab A (even when A was driven at constant speed) but was always accelerating and then decelerating. This is shown in Figure 8.5, where we have graphed the speed of both cabs versus time. Notice that there is a "wobble" in the case of cab B, even though A is moving at constant speed.

What would happen, we wondered, if there were more than two cabs in line. Figure 8.6 shows what happened when we added a third and then a fourth cab. The wobble was accentuated in the case of the third cab, while the fourth tended to accentuate it even more. Notice that the fourth cab goes into reverse; when we checked distances between cabs we also discovered that it had crashed through the cab in front of it.

We decided that the safe-distance approach was probably flawed. The algorithm, we hypothesized, led to *unstable* traffic flow in the sense that any slight fluctuation in the behavior of one cab was magnified in the behavior of the cab behind it. We tested this hypothesis for a number of values of the parameters *dt, k,* and *D* and never found a situation where a line of traffic moved safely and smoothly.

Did you reach a similar conclusion? Did you notice that any problems you had in getting one cab to follow another were magnified as you added more cabs to the line of traffic? Did you conclude that the engineers were wrong in their hypothesis that you could always avoid crashes by building a large enough safety factor into the distance D?

FIGURE 8.6
Simulating four cabs in line: A (moving at a constant speed of 20 m/s) in front, followed by B, C, and then D. This result shows how the wobble in speed is accentuated as we get further down the line. Even though the safe-distance algorithm is not always disastrous for cab B, it is bound to cause trouble somewhere down the line.

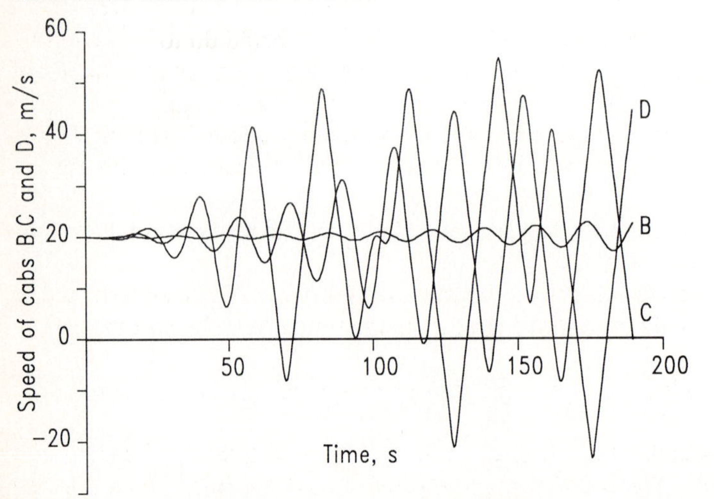

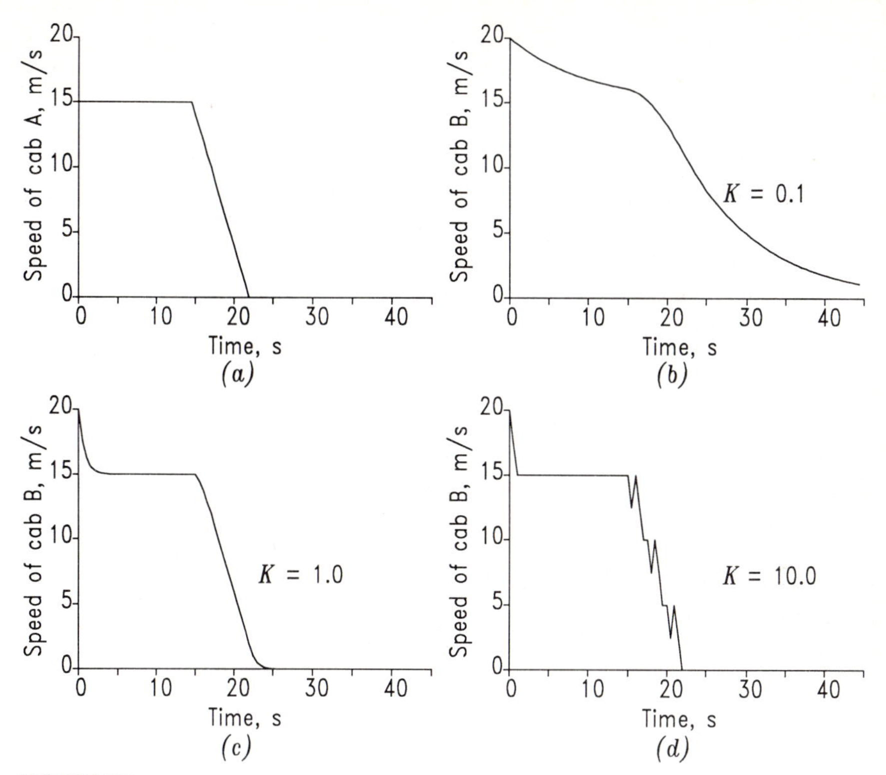

FIGURE 8.7
This set of graphs shows how cab B responds when cab A is decelerated: (*a*) shows the velocity of cab A as a function of time and (*b*), (*c*), and (*d*) show the response of cab B for different values of the parameter K in the speed-of-approach algorithm. $K = 1.0\ s^{-2}$ seems to give the best result.

Our Results: (2) The Speed-of-Approach Algorithm

Algorithm 3 behaved beautifully! There was little we could do to make it misbehave provided we chose values of dt that were not too large and values of the parameter K that were neither unreasonably small nor large.

We found that the braking test (where we applied one-half the maximum deceleration to cab A) was useful for tuning in the parameter K. Figure 8.7 illustrates this.

We then went on to test Algorithm 3 with four cabs instead of two. The traffic behaved well under all conditions.

Then we went on to predict the comfort of our passengers. Figure 8.8 shows the rate of change of acceleration of the cabs when the front cab is driven erratically.

Have you thought about what the rate of change of acceleration tells us?

It is commonly called the *jerk* and measures the smoothness of the ride. Figure 8.8 shows that Algorithm 3 actually provides cabs further down the line

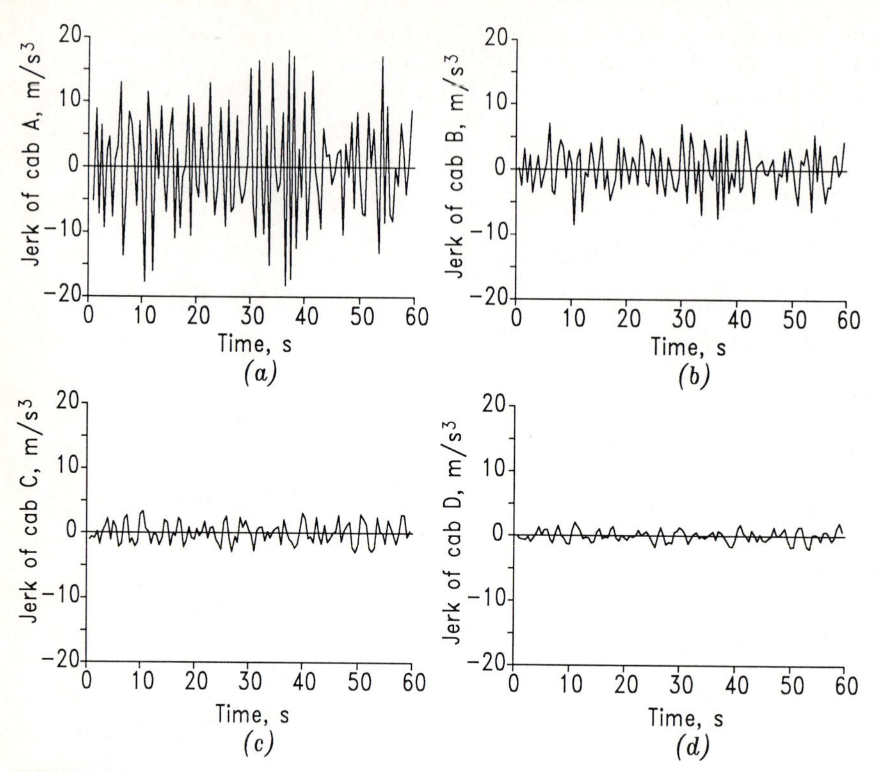

FIGURE 8.8
The jerk (or rate of change of acceleration) is shown for four cabs in line. Cab A is moving very erratically. Cabs B, C, and D follow, controlled by the speed-of-approach algorithm. Notice the comfort of the ride improves down the line.

with less jerky rides than the front cab. (Algorithm 2, incidentally, accentuated the jerk from one cab to the next.)

HOW WELL DID YOU PERFORM?

Did You Keep Your Algorithms Simple?

Simplicity is a relative concept; one can only discuss whether an algorithm (or model) is too simple or too complicated in terms of the stated objectives of the algorithm (or model).

There is always a tendency when one uses a computer to exploit its power, to make an algorithm just that little bit "smarter" without considering what this will achieve in terms of meeting objectives. It is a tendency that should be resisted. A good algorithm, like a good model, is one that does its job as parsimoniously as possible.

If you did make your algorithms more complicated, you probably discovered two important lessons:

1. The first is that the added complication is a nuisance to program. What looks relatively straightforward in a diagram or as a concept can lead to muddled computer code.

2. The second is that it is much harder to "tune in" a more complex algorithm and also much harder to interpret how it performs. In Algorithm 2, for example, if we assume that a_{max} and d_{max} cannot be changed, we still have three parameters that we can alter: the time interval *dt,* the safe distance *D,* and the constant *k*. It is possible, with only three "loose" parameters, to test thoroughly and conclude that the algorithm probably never performs well. However, each added complication will introduce at least one extra parameter.

How much extra work would be involved in thoroughly testing the behavior of a model with four parameters instead of three? Think about this carefully. And if you had five parameters?

On the other hand, one does not want to make an algorithm too simple. Algorithm 1 is an example of an algorithm that is too simple. In going from Algorithm 1 to Algorithm 2, we have introduced an additional parameter (*k*) but we can argue that the additional complexity is justified in terms of one of our objectives. We *thought* it would lead to a much smoother ride.

Also, simple algorithms often have implicit assumptions. It is important to recognize and evaluate those assumptions. For example, in all our algorithms we made sure that the maximum acceleration and deceleration were never exceeded. However, we never asked about the maximum *speed* that a cab could attain. There is a possibility that our control algorithms might try to make a cab go faster than it really can. You might have recognized this and *justifiably* made your algorithms more complicated.

We recognized this possibility only in retrospect (when our cab started reversing under the control of Algorithm 2), but even after recognizing it, we would not necessarily alter our algorithms. We prefer to argue that the cab leading the traffic (cab A) will never exceed its design speed and in fact will never even approach its design speed (remember the front cab is using the "open line" control routine). A good "following" control algorithm should never require the cab behind to drive much faster than the cab in front. Therefore an additional test to check on the speed of the cab is redundant. Our results with Algorithm 3 support this conclusion.

Do you agree with this argument?

Possibly not. You might want us to prove that it makes no difference. (In which case we would say to you: "That is a good idea, but a better idea is to ask *you* to go ahead and *disprove* it. This is the kind of experiment that your computer testbed was designed to accommodate.")

Alternatively, you might want to argue, from the safety point of view, that it makes sense to impose an upper limit on the speed of a cab. (Can you design a set of tests to demonstrate this? How would you "measure" safety?)

As you can see, it is not easy to decide which factors to include. What heuristics can one use for deciding whether or not an algorithm or model is too simple?

We recommend that you start simply, then progressively add complexity, and test to determine the usefulness of the additional factors.

Did You Plan Your Tests Carefully?

A model, once it is running reliably on a computer, is like a laboratory waiting to be used. Everybody knows that it is self-defeating to mess around in a laboratory without a clear objective or notion of what one is looking for, but not many people realize that it is equally self-defeating to run models aimlessly. The mere fact that the model is working seems to have a hypnotic effect: modelers cannot wait to see how their models perform. If you spent an inordinate amount of time running your model, without a clear idea of what was happening, then the chances are that you did not plan your tests with sufficient care.

The objective of the tests was to compare two algorithms. You cannot make comparisons if you have not identified precisely what it is you are going to compare. That is why we identified the three categories of safety, efficiency, and comfort and chose our computer output to provide data relating to each category.

There are other comparisons we could have made, and you might have investigated some of them. For example, there is a cost associated with accelerating and braking. Which algorithm provides the cheapest control? Or perhaps you recognized that there were better ways of monitoring efficiency, such as counting the number of cabs passing a point in a fixed period of time.

Another part of planning involves choosing plausible values for model parameters such as the safe distance D, the design parameters k and K, and the time interval dt. We were sloppy here. We decided we would manipulate these values and see what happened, and we paid the price by spending more time fiddling at the computer than we needed to.

Did you plan more carefully than we did? You might, for example, have estimated the safe distance D as follows:

Let V be the design speed for cabs in "following" mode. To determine the shortest distance in which a cab can stop, you could ask how far it will move if maximum deceleration is applied until the cab is at a complete stop. The answer is $V^2/2d_{max}$. *(Can you derive this formula?) You could then estimate D by adding a safety factor to this.*

Even our sloppy approach really involved some planning. We had to think about *plausible* ranges of values for the parameters and recognize that the choice of dt, for instance, was related to the choice of D. The argument is as follows. If D is 30 m and if cab B is approaching cab A at a relative speed of 10 m/s (which it might), then it will smash into cab A if the time step dt is not appreciably less than 3 seconds.

Choosing values for the parameters k and K was more like detective work. For example, the braking test (Figure 8.7) showed that the value of K had to be closer to 1.0 than to either 0.1 or 10. But why 1.0 rather that 3.0, for example? This is where the other tests played a part. If K is made somewhat larger than 1.0, then the algorithm does not damp out fluctuations. If K is made somewhat smaller, the chances of a crash during a braking test are increased.

What Did You Learn from Your Computer Runs?

We are pleased with how much we learned about controlling cabs from our computer runs. This was a modeling exercise that paid off handsomely.

Are you pleased with what you learned from your computer runs? How instructive was your model? Did you learn as much as you would have liked, or did you feel you spent far too much time "playing" unproductively with your model? If so, then perhaps you did not put enough effort into planning your computer runs or thinking about the results you obtained.

It requires experience (which is what we are trying to give you) and patience to plan your tests when the computer model is there just waiting to be run. But even after you have planned your tests you should not go to the computer straight away. You should learn to exercise even more patience and take additional time to *anticipate* your results. Having decided what tests you are going to run and what computer output you are going to plot, take an extra few minutes to think about how you expect that output to look. Draw rough sketches of the results you expect to get.

Did you think about what you expected the output to look like? If your results come out pretty much as you expected, then you can relax and say, "Yes, I think I understand what is happening." All you have to do then is look at the results in more detail and refine your understanding. However, if the computer results do not come out as you had expected, then you have to try to explain them, and this forces you to think about what is happening.

What you are doing when you anticipate your results is equivalent to stating a *hypothesis*. You are saying "This is what I believe." Your computer runs are then an experiment that supports, disproves, or leads you to modify your hypothesis.

Forming hypotheses is something you should also do while you are running your model. For example, the hypothesis that the safe-distance algorithm was unstable was not something we anticipated ahead of time. It was only while we were testing the algorithm that we noticed that small fluctuations in the speed of one cab led to larger fluctuations in the speed of the cab behind it.

Running a computer model successfully is thus like exploring a new world. When you start an expedition, you should be well prepared and you should have tried to anticipate what you will find. When you actually get there, you are also likely to want to explore interesting things you did not anticipate.

DISCUSSION

How well has this exercise illustrated the lessons we have drawn from it? Let us look at those lessons in more detail.

An Exercise in Simplicity

Throughout this exercise we urged you to keep your algorithms as simple as possible, and we practiced what we preached. The conclusions we drew from our algorithms illustrate very effectively how much one can learn from simple models.

But did we really convince you to keep your algorithms simple?

You probably learned more if you ignored our advice! Our experience is that most people are wary of simplicity. There is always the fear of having neglected something important. It is likely that the algorithms you devised were more complicated than ours and, with a bit of luck, you may have obstinately forged ahead and implemented them instead of pruning them. If you did, you probably have acquired a thorough appreciation of the advantages of simplicity. It must have been sobering, after all your hard work, to see how much we managed to deduce from our simple models.

It is easier for an instructor to let students learn this lesson the hard (but thorough) way in the classroom than for us to do so in a book. An instructor can casually ask "Are you keeping your model simple?" without actively pushing students to simplify. Often those who devise the most horribly complicated algorithms are exceptionally proud of their efforts and resist all suggestions that they should simplify what has taken so much effort to make complex. This is fortunate because they are then on the road to *discovering* how complexity handicaps them. Our experience is that people are always suspicious of the need to simplify until they have experienced it themselves.

However, the lesson to be learned about simplicity is not quite so simple after all. We stressed earlier that simplicity is relative; it depends on objectives. This is a rich problem for illustrating this point. All you have to do is change your objectives and ask whether your algorithm is still appropriately simple.

For example, suppose an important objective is to increase the flow of traffic. Will Algorithm 3 still be a good choice?

We suspect not. It matches speeds, but does not control the distance between cabs. To increase the flow of traffic you want to decrease the distance between cabs. This objective therefore probably forces you into making the algorithm more complex.

How would you modify it?

Learning to Use Models Effectively

This was also an exercise in how to *use* a model purposefully and thoughtfully.

Did you feel that the problem had been "solved" as soon as your model ran without "bugs" on the computer? Did you just produce some typical output and write a report? Or did you go to the other extreme and produce a report with page after page of computer-generated graphs?

The main intellectual effort in this exercise was in testing the two algorithms, interpreting the results, and drawing conclusions. That intellectual effort is provided by you, the modeler, not the computer. It is therefore important to encourage habits that force you to *think* about what you are doing whenever you run your model. The habits we suggest you cultivate are:

- Plan a series of experiments on the computer in the same way that you would plan experiments in a laboratory.
- Think about what output you need, what you will do with it, and how you expect it to look, i.e., anticipate your results.
- If actual results do not agree with anticipated results, make sure that you understand the reason.
- Postulate hypotheses and think of your computer runs as experiments to test the hypotheses.
- Look carefully at your results for unexpected trends.
- Plan well, but also be flexible and be prepared to postulate new hypotheses during the course of your runs.
- Do not think of the computer output as the end product of your runs. The output is a means to an end, not an end in itself. The purpose of generating output is to draw conclusions. It is therefore essential to ask yourself questions such as "What have I discovered?" or "Can I summarize the output in words?"
- When you have answered those questions, look again at your output and think about how to illustrate the conclusions you have drawn. That will determine which of your runs or graphs to use in your report.

Reiterating Lessons from Previous Chapters

We introduced our notation for the speed and position of both cabs A and B without much of a fanfare, but notice how important it was to develop an appropriate notation. The discussion of an algorithm or description of a model is made that much easier and clearer when it is based on good notation. The question "Have you identified the important variables and introduced a useful notation?" is one you should always ask early on. You should also keep on asking it over and over again.

Notice how putting you to the task of explicitly describing the control algorithms encouraged you to think about notation.

Notice too how useful it was to try to draw a diagram that represented your approach to modeling the cabs. It is often easy to persuade yourself that you understand what you are doing when in fact your ideas are only half-baked. Writing something out explicitly or drawing an unambiguous diagram forces you into recognizing when your ideas are half-baked and encourages you to "bake" them.

A good heuristic is: If you think you know what you are doing, ask yourself a question or set yourself a task to convince yourself that you know what you are doing!

The Concept of Stability

We have used the word "stability" a little loosely when we have talked about the stability of a line of traffic. What we mean by it is that any quirks in the behavior of one cab should be reduced, not magnified, by the cabs behind it. The safe-distance algorithm magnifies fluctuations in speed, which is why we called it unstable. The fact that fluctuations are magnified down a line of traffic is not a problem in itself, but it becomes a problem when the fluctuations grow to be so large that they endanger the flow of traffic.

The concept of the *stability of an algorithm* is important in computing. Whenever a computer uses a formula to calculate a real (noninteger) arithmetic expression, it makes small errors. These errors are trivial in themselves, but if the answers are fed back into the same formula, and the errors are then magnified, eventually they may lead to ridiculous results.

To demonstrate this, you may want to consider the following rather contrived but illustrative exercise:

Suppose you want to generate the sequence

$$3, 1, 1/3, 1/9, 1/27, 1/81, \ldots$$

on a computer. Consider the following three formulae:

1. Set $x_{n+1} = x_n/3$, given $x_1 = 3$

2. Set $x_{n+2} = (301\, x_{n+1} - 100\, x_n)/3$, given $x_1 = 3$ and $x_2 = 1$

3. Set $x_{n+2} = (13\, x_{n+1} - x_n)/30$, given $x_1 = 3$ and $x_2 = 1$

You can check by hand that all three formulae should produce the required sequence. But that is not what happens if you implement them on a computer. Try producing the first 20 terms of the sequence using these formulae on a spreadsheet or your own program. What do you notice?

The first and last formulae behave well. The middle formula starts off well, but eventually deteriorates and produces utter nonsense. The reason for this is not that it is a clumsy formula (the last formula is equally clumsy) but that it is an *unstable* formula. It magnifies errors, whereas the last formula reduces errors.

The safe-distance algorithm is unstable in a similar sense: it amplifies any small wobble in the behavior of a cab down the line of traffic until cabs are behaving totally erratically. In contrast, the speed-of-approach algorithm damps down fluctuations.

DIFFERENTIAL EQUATIONS
(For those with a background in Calculus)

Consider Algorithm 3. If we ignore the maximum acceleration and deceleration, then we can write

$$a_n = K(u_n - v_n)$$

and the speed of cab B at time $(n + 1)dt$ becomes

$$v_{n+1} = v_n + K(u_n - v_n)dt$$

If we write this in the form

$$\frac{(v_{n+1} - v_n)}{dt} = Ku_n - Kv_n$$

then in the limit as dt tends to zero, the left-hand side is just the derivative dv/dt. As dt becomes arbitrarily small, we therefore have

$$\frac{dv}{dt} = Ku(t) - Kv(t) \qquad (8.5)$$

This is a *differential equation*. If we specify $u(t)$, which is the speed of the cab A, it can be solved to give us $v(t)$, the speed of cab B. For example, if cab A is moving at a constant speed U, then

$$\frac{dv}{dt} + Kv = KU$$

If you have taken a course in differential equations, you should be able to solve this to get

$$v(t) = U + Ce^{-Kt} \tag{8.6}$$

where C is a constant that depends on the speed of cab B at time $t = 0$, i.e., it depends on how fast cab B was moving when it first switched from "open line" to "following" mode.

Notice that the actual value of C is unimportant because the exponential term dies out with time. Equation 8.6 tells us that cab B eventually matches the speed of cab A.

What happens if the speed of cab A fluctuates? Suppose, for example, its speed is $U + c \sin(wt)$, where c and w are constants. Then the differential Equation 8.5 becomes

$$\frac{dv}{dt} + Kv = KU + Kc \sin(wt)$$

This is harder to solve, but the solution can be found and is

$$v(t) = U + Ce^{-Kt} + Kc\,\frac{K \sin(wt) - w \cos(wt)}{K^2 + w^2}$$

which can be rewritten in the form

$$v(t) = U + Ce^{-Kt} + \frac{Kc}{(K^2 + w^2)^{1/2}} \sin(wt + b) \tag{8.7}$$

where b is a constant called the phase angle.

Notice that this tells us that if cab A has a speed wobble, so will cab B. Moreover, cab B will have a wobble that is out of phase with cab A; but, more importantly, since

$$\frac{Kc}{(K^2 + w^2)^{1/2}}$$

is *always* less than c, the amplitude of cab B's wobble will be *less* than the amplitude of cab A's wobble. Any wobbles will therefore *die out* in a line of traffic. This is in direct contrast to our experience with Algorithm 2, where wobbles were amplified along a line of traffic. (One can prove this using differential equations too.)

This analysis therefore underscores why Algorithm 3 is stable. It is, however, worth pausing to think about what we have done, from the modeling point of view, in reducing the algorithm to a differential equation.

When we write down a differential equation, our representation of *time* changes. The equations we used previously (with subscripted notation) are called *difference equations*. In a difference equation, time "jumps" by a discrete amount (*dt* in this case) from one step to the next. Models built using difference equations are therefore called *discrete* models. On the other hand, time flows smoothly and continuously when we build differential equation models; we therefore talk about *continuous models* in that case.

This can be an important distinction. Just as in previous chapters we paused to consider whether to build a deterministic or stochastic model, so we should also deliberately decide whether to build discrete or continuous models. However, this decision depends in part on your mathematical background. If your calculus skills are inadequate, there is little to be gained by building a continuous model, and you can "fudge" it by choosing a very small time step in your discrete model.

Think of previous chapters in this book. Where would a continuous model have been more appropriate than a discrete model?

In both the gas tank problem of Chapter 3 and the beer cooling problem of Chapter 4. Right? In both those cases we did indeed "fudge" the solution by choosing very small time steps. There is nothing wrong with this, but we do have to be aware of the fact that our model is a discrete representation of a continuous process. In fact, in many scientific and engineering applications, discrete computer solutions are used as *approximations* to differential equations. (Have you ever solved a differential equation numerically?)

Getting back to the cab-following algorithm, which is the better model in this case: a discrete model or a continuous model?

Here the differential equation is the "fudge." It is only an approximation of the control algorithm! It is an approximation for two reasons:

- In practice the time interval *dt* may be small, but not infinitesimally small; it does not tend to zero.
- Equation 8.5 does not correct the acceleration or deceleration if either exceeds the maximum.

Equation 8.5 is therefore a simplified, stripped-down version of the control algorithm. It has certain features in common with the control algorithm, but is not identical to it. We can therefore think of it as a *model* of the control algorithm. It is a useful model because it leads to mathematical expressions such as equation 8.7 which are easy to interpret and are suggestive about the behavior of the actual algorithm.

That, after all, is what modeling is all about!

SIMILAR PROBLEMS AND FURTHER READING

In this chapter we have developed quite a complex dynamic (things are moving or changing with time) model. You will find the formulation and solution of many more dynamic models in *Learning the Art of Mathematical Modeling* by Mark Cross and A. P. Moscardini (New York, Wiley, 1985).

An important concept is that of stability in the sense that small errors or fluctuations introduced in a process (such as a line of traffic) do not grow out of control. The following two problems illustrate this or similar concepts.

A Medical Problem

Certain bacteria are capable, if their numbers increase sufficiently, of causing a bladder infection. Doctors have two possible remedies: to introduce an antibiotic that attacks and kills the bacteria or to increase the patient's throughput of fluids.

Consider bacteria that occasionally enter the bladder. Suppose that they are capable of doubling their numbers every 30 or 40 minutes. Develop a model that would enable you to predict whether or not the bacteria would be likely to cause an infection. Use your model to investigate the two alternative remedies.

Are you going to build a deterministic or stochastic model?

We suggest that you develop a deterministic model. After you have drawn conclusions from it, however, it could be instructive and interesting to see how well those conclusions stand up if you make your model stochastic.

A Fishing Problem

A large lake contains two delicious species of fish, known to local fishing enthusiasts as *glubs* and *slobs*. Slobs eat glubs; in fact glubs are the main component of their diet.

A research team from a neighboring university has studied the interaction between the two species and has even built a model to describe it. If G_n represents the glub population in year n, and S_n the slob population that same year, their model can be written

$$G_{n+1} = G_n + aG_n - bG_nS_n - cG_nG_n$$

and

$$S_{n+1} = S_n - pS_n + qS_nG_n$$

Look carefully at these equations, and write them down in words. Can you explain what each term represents?

The research team's best estimates of the parameters in the model are

$$a = 0.15$$
$$b = 0.75 \times 10^{-3} \text{ per slob}$$
$$c = 0.30 \times 10^{-5} \text{ per glub}$$
$$p = 1.5$$

and

$$q = 0.60 \times 10^{-4} \text{ per glub}$$

If the two species are in equilibrium, how many of each species would you expect to find in the lake?

There is a proposal to harvest glubs and/or slobs commercially. This commercial fishery is being opposed by environmental groups, and as a compromise a quota system has been proposed. In terms of this quota, the fishery will be allowed to remove only a fraction (r_G) of the glubs and a different fraction (r_S) of the slobs in the lake in any year.

You have been hired by the fishery. Their economist tells you, in confidence, that the profit on a slob is exactly ten times the profit on a glub. Your task is to suggest values for r_G and r_S that will ensure the best long-term sustained profitability.

CHAPTER 9

THE CASE OF THE DISHONEST ADVERTISER

BACKGROUND

For some months Z had been running the following advertisement in the local newspaper:

> Disorganized executive is anxious to pay a fabulous salary to a superbly qualified personal secretary. Willing to enter into a generous, long-term contract with the right person. Write to P.O. Box 007 if you wish to be interviewed.

In fact, Z had no intention of paying the kind of salary the advertisement suggested. Each month he carefully chose the best-qualified applicant and said, "You look very promising indeed, but I am sure you will agree that it is only fair to both of us that we should have a trial period of a month before signing a long-term contract." Needless to say, the salary he paid for that month was not even competitive, and at the end of the month he found an excuse to terminate employment. He then recruited a new secretary via his advertisement. In this way, he secured top-rate secretarial service at minimal cost.

Eventually Z was reported to the police and was charged with fraud, dishonest advertising, and a number of lesser crimes. The judge who heard the case found Z guilty of all charges but had an old-fashioned sense of justice. He sentenced Z as follows:

> You advertised that you intend to offer a permanent and attractive appointment, so it is only just that you should carry out your stated intent.
>
> I have confiscated the contents of your post office box and found in it eight unopened responses to your advertisement. I contacted each of these applicants and have ascertained that they are all serious and willing to work for you. Exactly two

weeks from today you will report to my chambers. There you will interview the eight applicants. I will organize the interviews at 15-minute intervals, and you will only meet one applicant at a time. At the end of the first interview you have to make a choice: either you hire the first applicant (without seeing the others), or else you say "no." If you say "no," you will never see that applicant again, but will have the same choice at the second interview, that is you can offer employment to the second applicant or say "no" irrevocably and go on to interview the third applicant. If you continue to say "no" to the first seven applicants, then you have no choice but to employ the eighth and last applicant.

My clerk will inform you of the terms of employment you will be obliged to offer the successful applicant. Suffice to say that there will be no trial period, that the contract will bind you for twenty years, and that the salary will be every bit as fabulous as your advertisement suggests.

YOUR PROBLEM

You too have been advertising in the local newspaper. Your advertisement reads:

> Professional problem solver (with personal computer) seeks challenging assignments.

Z has just visited your office and told you the above story. He is a reformed character. His objective, he explains, is to make the most of an unfortunate situation. If he has to employ a secretary for twenty years on exceptionally generous terms, he wants to do his utmost to make a good choice. Hence his visit to you. He wants you to advise him on when to say "yes" to one of the eight applicants.

Z is somewhat skeptical about newspaper advertisements. He is unwilling to hire you outright as a consultant. You have therefore been retained, in the first instance, to have a preliminary look at the problem and make a presentation to Z on how you would tackle it. As Z says, "I want you to tell me what you can do to help me."

Since time is short, you have an appointment with Z for tomorrow. Your first task, therefore, is to think carefully about how you as a modeler and problem solver can advise Z.

Make a list of questions to ask Z, and then prepare your presentation. The presentation need not go into details; think of it as a "game plan" that outlines the main features of your approach. Remember that you have a busy schedule and probably cannot spare more than an hour on this assignment.

✔

QUESTIONS TO CONSIDER

What information do you and Z have? What else could Z possibly tell you? Can you make use of everything you know? What further information will Z have at the time of the interviews? Can you use this additional information?

What assumptions can you make? Are they reasonable? How will they help you?

Have you thought back to the heuristics you used in previous problems? Are any of them useful here? For instance, can you think of an easier version of this problem that would help you structure your thoughts?

Suppose Z had access to a phone during the interviews and wanted to consult you. How would Z describe the applicants to you? What kind of description would be useful to you?

Have you thought about the kind of advice that you can give to Z? Would it be appropriate, for instance, for you to say "Hire the third applicant"? How much of a gamble would it be if Z followed that advice? Could your advice be worse? How could you give better advice?

We have used the word "gamble" in the previous question. Have you recognized the element of chance in this problem? Have you thought about how this affects your approach to the problem? Is it useful to think of other gambling situations and how you would analyze them?

If, after reading these questions, you are happy with your presentation for Z, go ahead to the next section. If not, spend some more time in preparing your presentation.

✔

SOME ANSWERS TO THESE QUESTIONS

There are different ways in which one might answer the above questions. In this section we will present our answers to give you some idea of how we would think about Z's problem. We do not expect you to think in the same way, but we hope that in the process of answering the questions we will be able to identify *key* issues that *any* approach to the problem should address.

Information and Assumptions

What information do we have?

The only incontrovertible fact is that there are eight applicants. Apart from knowing that they are all prepared to work for Z, we know nothing more about them.

What can Z tell us?

Presumably Z is anxious to employ the most suitable applicant, but we have no idea of what "suitable" means to Z. An important question for Z is "What sort of secretary are you looking for?"

What assumptions can we make?

Perhaps a better question to ask first is "Where do we need to make assumptions?" Usually we have too much information and make assumptions about what we can ignore or neglect. We certainly do not have too much information in this problem; we have too little. Perhaps then we need to make assumptions about the information we do not have.

Where, then, are we really hurting for lack of information?

The answer to that question must surely be that we would like to know more about the eight applicants.

What, in particular, would we like to know?

First, we would like to know something about the *order* in which Z will interview them, and, second, we would like to know something about their qualities: How do they match up to Z's wish list for the perfect secretary?

We do not know the answers to these questions, but can we make any reasonable assumptions?

The judge spoke to the applicants, but presumably knows no more about them than we do. Can we therefore assume that he has not scheduled them in any deliberate or diabolical order? The answer to this is probably "yes," and so our first assumption is that the order of interviews is *random*.

How does this help us?

It would have been far more useful if we had known that there was a deliberate pattern. In fact the solution to Z's problem would have been trivial if we had known that the "best" applicants were going to be scheduled for the early interviews. On the other hand, if there was a secret deliberate pattern and we had no way of finding it out, we would have been worse off rather than better off. It helps, therefore, to know that there is no pattern. We can think of the eight names going into a hat and being pulled out by a blindfolded judge. We might even ask Z's lawyer to insist that the order of interviews be scheduled in this way.

We also know that the judge found the applications in Z's post office box. Presumably, therefore, they are no different in any important way from all the applicants that Z has met in the past. This does not tell us anything specific about the eight applicants, but in general it suggests that we can draw on Z's past experience. This is our second assumption.

What additional information will Z have at the time of the interview?

Z will know how many applicants have been refused (and therefore how many are still to come). He will also know something about the qualities of the applicants he has turned down as well as the qualities of the applicant currently being interviewed.

We may not be able to use all the above information, but it is all potentially relevant, and the quality of the advice we give Z will depend on how much of this information we take into account and how well we use it.

Our presentation to Z would therefore begin with "This is what we know and this is what we will assume" and would review the above points.

Looking for a Simpler Problem

Is there a simpler version of this problem that we could think about?

We wondered whether it was useful to consider how we would solve the problem if Z were allowed to interview all the applicants at the same time.

This is a simpler version of the problem. It highlights the need to quantify, or at least rank, the qualities of the applicants. It brings us back to the question "What is Z looking for in a secretary?" This is clearly something we need to discuss with Z, but an idle discussion will not be sufficient. We need to establish clear-cut criteria that tell us that applicant X is preferable to applicant Y. We would therefore ask Z, on the basis of past experience, to think about how to rank the applicants.

Notice that we are already making use of one of our assumptions: we are drawing on Z's past experience.

Asking how Z could usefully describe the applicants to us over the phone also highlights the need for some sort of ranking. Notice that we are not interested in the specific qualities of each applicant, only in a comparative ranking.

We begin to see that the key issues during the interviews will be which applicant Z is currently interviewing (is it the first or the fifth, for instance?) and how that particular applicant ranks.

What Kind of Advice Can We Give?

It really makes no sense at all to say "Hire the third applicant." It does make sense, if we have agreed with Z on a system for ranking applicants, to give advice such as "Hire the first top-ranking applicant you interview."

How much of a gamble would it be if Z were to follow this advice? In other words, what risk does Z take by hiring the first top-ranking applicant?

If Z meets a top-ranking applicant, then obviously Z will be happy. The risk that Z takes is that none of the eight applicants may be top-ranking, in which case Z will have to employ the eighth applicant (who might be most unsuitable).

If we carry this thinking a little further, we should ask ourselves what information we need to calculate the risk that none of the applicants will be top-ranking?

That depends on whether top-ranking applicants are rarely or less rarely encountered. This leads us back to the assumption that we can draw on Z's past experience. We need to ask Z questions like "What percentage of previous applicants were top-ranking?"

We begin to see the structure of an approach to Z's problem:

- We must find a way of *ranking* applicants.
- We must suggest *strategies* for making decisions at the interviews.
- We must find a way of estimating the *risks* associated with each strategy.

The Element of Chance

Have we thought about how the element of chance affects our approach to this problem? We have used words like "risk," but we have not really come to grips with how to cope with chance. How does it limit us? If we build a model, will it be stochastic or deterministic?

By using the word "risk" we have admitted that we cannot tell Z the inevitable outcome of following our advice. We cannot say "If you follow this strategy, then this is what will happen." We can say "If you follow this strategy, then this is what *might* happen." That is not very useful to Z unless we can quantify the word "might."

Can you think of how one could do that? How can we express the likely outcome of following one strategy rather than another?

We could, for instance, tell Z that "If your strategy is to hire the first top-ranking applicant, the chances are 82 in 100 that you will in fact find a top-ranking applicant. The chances are also 7 in a 100 that you will be forced to employ a very low ranking applicant." This is the kind of information that we can only get out of a stochastic model. Our approach to this problem will therefore be similar to the way we modeled the tennis problem of Chapter 5.

Thinking of the way we tackled the problem of betting on tennis games, or how we might tackle any game of chance, such as blackjack, is helpful. If a blackjack player holds cards totaling 16 points and is considering whether to take another card, it would be helpful to know the risks involved. What is the *probability* that the additional card will have a score of more than 5? If we knew something about the cards that had already been played, we could estimate it. The blackjack player's decision is not all that different from the decision Z faces at each interview.

What kind of model could we build? How would we use it?

Z will experience the interview process only once. Because of the element of chance, there is very little that we can say about that one set of interviews. But suppose we had 100 or, better still, 1000 clients identical to Z and we gave them all the same advice. If we knew what happened to each of them (i.e., who they employed), we could draw conclusions such as "27 in 1000 had to employ applicants in the lowest rank." In other words, the probability of having to employ a candidate in the lowest rank (as a result of following our advice) is estimated to be .027. This would enable us to *quantify* the risks associated with our advice.

It would be useful to build a model that generates hundreds of realistic sets of interviews on the computer. We could then use these sets as data to calculate the risks associated with different strategies.

It is too early to go into the details of such a model, but at least we have a structured approach to present to Z.

How do our ideas compare with yours? Does your approach contain the key concepts of

- *A way of ranking applicants?*
- *A strategy?*
- *Risk?*
- *A model for relating strategy to risk?*

Does your list of questions for Z include questions that explore how Z will rank the applicants? Does it also include questions that relate to Z's experience of previous applicants?

In the next section we present a conversation you might have had with Z if you had tried to ask these questions.

AN IMAGINARY CONVERSATION WITH Z

You: What sort of secretary are you looking for?

Z: The perfect secretary, of course!

You: Come on, that's a little trite. What are your criteria for a good secretary?

Z: Well,...somebody fluent in French and German.

You: If you couldn't find somebody fluent in both languages, which would be your first choice?

Z: French, I guess, but German is important too. Also, I'd like somebody who is good with figures: bookkeeping and tax and that kind of thing.

You: What is more important to you, strength in languages or a background in accounting?

Z: Well, both really. Other things are intelligence, somebody who can take the initiative, be tactful with clients, preferably a good typist too....

You: Hang on a minute. Can you put those in order? If you had to choose between typing and tact, for instance....

Z: ...Oh, and punctual, I must have a secretary who is punctual. At the salary I'll be paying I want the best: loyalty, a sense of humor, and so on. In fact, a "ten" on a scale of one to ten!

You: How often have you met a "ten"? That is, among the previous respondents to your advertisement?

Z: Actually, never!

THE NEXT STEP

Despite this somewhat unsatisfactory conversation, Z likes your approach. He has hired you to go ahead and fill in the details. Z wants to know exactly how you intend to

1. Develop the scheme for ranking applicants
2. Compute the risks associated with different strategies

You will be meeting again in 2 days' time. Plan your model in detail now. If you have the time, go ahead and implement it. Z would really be impressed if you brought along a sample output.

If you are confused after the conversation with Z, read ahead, but preferably only after thinking about what you would do next. We urge you to make assumptions. Pretend that you have access to Z; if you need to ask him a question, invent the answer you think he might have given.

✔

THE CONVERSATION WITH Z IN RETROSPECT

Before we had our imaginary conversation with Z, we had a fairly clear idea of the structure of our model. The conversation with Z has muddied the waters. Whereas previously we had too little information relating to Z's problem, we now have a wealth of confusing detail. If we were to continue questioning Z, it is likely that the details would accumulate; we would be confused rather than helped.

This is not an unusual experience. Problem solvers are often confronted with more details than they can handle. What can we do under these circumstances?

We only have three options:

1. Go back and restructure our model, taking into account what we have learned from our conversation with Z.
2. Include the new details in our model without restructuring it.
3. Somehow cut through the details and forge ahead with our original ideas.

Which option should we take? Think about this before you read on. (To make a good choice you need to remember what you were trying to find out from your conversation with Z.)

✓

We had two objectives in our conversation with Z: we wanted to find out how Z ranked applicants, and we wanted to delve into his experience of previous applicants. Nothing Z has told us suggests that there was anything wrong with these objectives. Z's answers are merely more complicated than we might have wished. This suggests that we can ignore option 1; there is no need to restructure our model.

If we do not have to restructure our model, it makes sense to consider option 3 before we consider option 2.

Our heuristic is: Always try a simpler approach first, then refine as necessary.

The basic problem is that we want Z to rank the applicants unambiguously on a single scale, whereas Z has a number of scales (languages, accounting, typing, intelligence, etc.). How can we persuade Z to use only one scale?

Z has in fact given us a clue: remember that Z wanted to find a "ten" on a scale of one-to-ten. We should find out whether Z can *subjectively* rank applicants on that scale without worrying about the details. The chances are that Z can. Alternatively, if Z finds it too difficult to use the one-to-ten scale, we could suggest something simpler such as a four-category scale of "poor," "marginal," "average," and "good."

If Z cannot do this, then we have to consider option 2. In effect, we have to help Z map a number of criteria onto a single one-to-ten or poor-to-good scale. This would be a task in itself, a model within our model. It is not a task that we will pursue, although you might like to think about how you could do it. For the purposes of this exercise it suffices that, one way or another, we are going to use a single ranking scale.

There is still the problem of relating this scale to Z's past experience. We know, for instance, that Z has never met a "ten" among the previous applicants.

How can we summarize Z's past experience in a form that will be useful for our model?

If we are using the poor-to-good scale we could ask Z to estimate the percentage of previous applicants that he would have categorized as "poor," the percentage that were "marginal," and so on. We could do something similar on the one-to-ten scale.

Why is this likely to be useful? If you used the one-to-ten scale, how would you cope with the fact that Z had never met a "ten"?

COMPUTING RISK

In the previous section we outlined how we would develop our scale for ranking applicants. You might have other ideas; provided that your ranking procedure is unambiguous and can be related to Z's past experience, it is likely to serve its purpose. However, clarifying how you would rank applicants is only the first of your tasks for the next meeting with Z.

The other task is computing the risks associated with whatever strategies Z might consider. Go ahead and build a model that does this (if you have not already done so), using either your ranking procedure or one of our scales. If you need guidance, read the next section.

✔

COMPUTING RISK: POINTS TO CONSIDER

What is your overall strategy for computing risk?
How will you assign a rating to the eight applicants in one set of interviews?
What will the output from this look like?
How will you use Z's scale for rating applicants?
How will you draw on Z's past experience?
How many sets of interviews are you going to have to simulate for each strategy?
How are you going to represent risk?
How will you present the results to Z?

If your model addresses these points, read on and compare it with ours. Otherwise spend some more time thinking about your model.

OUR MODEL FOR COMPUTING RISK

Suppose we have chosen a scale for ranking the applicants. Suppose too that we have chosen a plausible strategy. To compute risk, we must build a model that will tell us what might happen if we applied that strategy over and over again. That means that we want to simulate *many* sets of interviews. For each set, we want to:

- Assign a ranking to each of eight applicants, in order

• Apply the strategy to those eight applicants and record the ranking of the applicant Z would hire using that strategy

From the recorded results of a large number of simulated sets of interviews, we can then compute probabilities that enable Z to assess the risks associated with the strategy we are testing.

Let us look at how we would do this in detail.

Step 1. Suppose that Z's scale consists of the categories "poor," "marginal," "average," and "good."

Step 2. Suppose that our strategy is to employ the first "good" applicant.

Step 3. We want to assign a category to each applicant. Table 9.1, for example, shows how we might end up describing the eight candidates. In this case we have somehow decided that the first candidate is "poor," the second "good," and so on.

The question you should ask is "What distinguishes Table 9.1 from any other set of eight categories that we may have chosen to write down?"

The answer must surely be that Table 9.1 is useful if it is representative of Z's past experience; otherwise it serves no purpose whatsoever.

We decided earlier to ask Z questions such as: "What percentage of applicants, in your experience, would you have categorized as 'poor'?" and "What percentage were 'marginal'?" and so on.

Suppose Z answered, "Two out of six were 'poor,' one out of six was 'marginal,' two were 'average' and only one in six was 'good.'" Our simulation will be representative of Z's experience if we assign categories at random but in such a way that on average two out of six are "poor," and so on.

This should remind you of how we decided who would win a point in the tennis problem of Chapter 5. In statistical terms, the information we have

TABLE 9.1
RANKINGS GENERATED BY A COMPUTER ALGORITHM

Candidate	Ranking
1	Poor
2	Good
3	Marginal
4	Average
5	Poor
6	Average
7	Poor
8	Marginal

about Z's past experience defines the *distribution* of the four categories among the population of applicants. We want to assign categories in our simulation at random, but in such a way that they have an identical distribution.

It is easy to see how we could do this with a die instead of a computer. We could throw the die eight times. Our algorithm might be:

- If numbers 1 or 2 come up, write down "poor."
- If number 3 comes up, write down "marginal."
- If numbers 4 or 5 come up, write down "average."
- If number 6 comes up, write down "good."

You are probably ahead of us at this stage and have already worked out how you will use a random number generator to assign categories. If not, then the analogy of the die should give you a clue.

Table 9.1 was generated using a random number generator instead of a die. It satisfies the two essential criteria: it is generated randomly and it belongs to the same distribution as Z's past experience. But we cannot tell this just by looking at it. If we were to repeat the exercise, we are likely to get a completely different table. We can only tell whether we have used Z's distribution by generating many such tables and then checking to see whether approximately one-sixth of all the entries are "good," and so on. This underscores why we need to simulate many sets of interviews in order to compute risk.

Step 4. By looking at Table 9.1, we see that the strategy in step 2 would lead to Z employing a "good" applicant in this case. The result for this interview is therefore "good."

Step 5. We then repeat steps 3 and 4 a large number of times. Suppose we decide to repeat the steps 500 times. We will keep running totals of the results, i.e., we will record how many times Z employs a "poor" applicant, and so on. At the end of the 500 runs we can convert these totals to probabilities. We could then tell Z the probability of hiring each of the four categories as a result of following this strategy. All this information is useful to Z. We might present it in a histogram, such as Figure 9.1.

How do we know that repeating steps 3 and 4 five-hundred times is sufficient?

As in the tennis problem, we could consult a statistician, or we could experiment on the computer to see how many replicates we need to confine the fluctuations in our results (the fluctuations in the probabilities we calculate) to within a narrow band.

We looked at the fluctuations in the probability of hiring a "poor" candidate and concluded that 500 replicates were indeed sufficient for

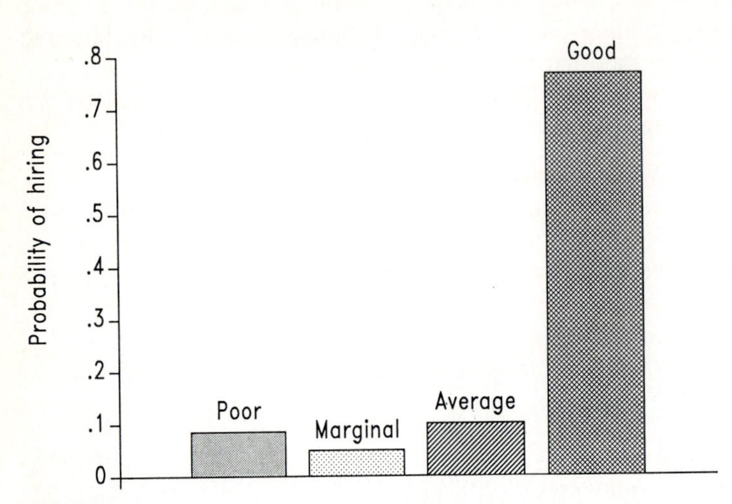

FIGURE 9.1
An example of how you might want to present the results from your model. This diagram shows the probability of hiring a secretary in each of four categories. This histogram was compiled from 500 replicates. There were eight candidates in each case, their rankings were generated from a given distribution, and the same strategy for choosing a secretary was applied in each of the replicates.

our purposes. (Why do you think we chose to look at fluctuations in the "poor" result rather than in, for example, the "good" result?)

Figure 9.1, plus an explanation of what we were trying to achieve, is what we would present to Z at our next meeting. We may choose a scale of one to ten instead of the poor-to-good scale, but the principle will be exactly the same.

Did you develop a similar model? In particular, did you draw on Z's past experience? Notice that the model is meaningless without that experience.

EXERCISING YOUR MODEL

Assume Z likes your model. Your final task is to use it to test out three or four likely strategies and to present Z with an assessment of the risks associated with them.

Before you go ahead and do this, you should consider the following points:

1. What strategies are you going to test?
2. On what basis will you recommend one strategy rather than another?

✔

OUR SOLUTION

We decided to choose three strategies: one optimistic, one pessimistic, and one somewhere in between. In this way we hoped to bracket the possible range of strategies, see how Z reacted to them, and then, if necessary, look at more complex strategies with a clearer idea of what Z is looking for.

- Our optimistic strategy is: choose the first "good" applicant.
- The pessimistic strategy is: hire the first "good" or "average" applicant.
- The strategy in between the two is to hire only "good" applicants for the first five interviews, but accept either a "good" or "average" applicant thereafter.

We tested each strategy 500 times. The results of those tests are shown in Figure 9.2.

We made no attempt to persuade Z to choose one strategy rather than another. After all, Z has to live with the decision for 20 years, not us. We did, however, explain to Z how the histograms in Figure 9.2 could be used to assess trade-offs. For example, if Z is terrified of hiring a "poor" candidate, then by comparing the optimistic and pessimistic strategies, Z can see that the chances of getting a "poor" secretary are reduced drastically in the pessimistic strategy, but at a price—the probability of finding a "good" secretary is reduced too: from .77 to .34. This information should help Z decide just how cautious he wants to be.

FIGURE 9.2
The same type of histogram as Figure 9.1 but this time showing the results for three explicit strategies (as described in the text). Notice how these histograms illustrate the trade-off between high expectations and the probability of hiring a poor or marginal secretary.

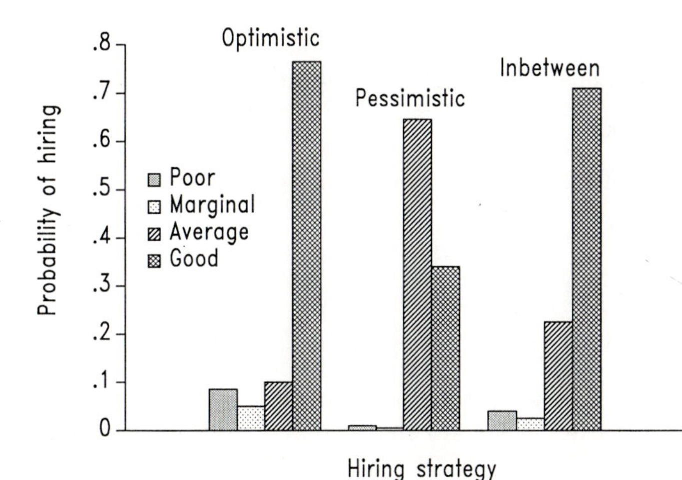

DISCUSSION OF THE MODEL

Now that you have seen our model and solution to Z's problem, compare it with yours. Where do you think your model is better than ours? Where do you think our model is better than yours?

There are a number of points to be made about what we did and also about what you might have done:

Choosing a Scale

Choosing a scale for comparing applicants was the key to drawing on Z's past experience and providing a basis for analyzing his present dilemma.

Why did we insist on using only one scale? After all, we know that Z is looking for a number of different attributes.

We chose one scale in order to simplify comparisons. How do you compare two applicants if one scores higher than the other on one scale, but vice versa on a second scale? You would have to somehow weight the different attributes that the two scales measure. This could be difficult to do, but we have cut through the problem by assuming that Z is capable of reducing preferences to one scale only.

Does this remind you of the Gordian knot?

If Z were not capable of doing this, or if Z really needed our help in doing it, then we would have been faced with an additional problem: that of coming up with a formula or algorithm that would have taken into account all the attributes that Z felt were important. We would have had to weight them in a way that satisfied Z. This is the model within a model that we mentioned earlier and decided to ignore.

Having decided to use a single scale, we next had to choose between the poor-to-good scale and the one-to-ten scale.

Obviously there are other scales one might use, but of these two, which is preferable?

We have chosen the poor-to-good scale because it simplifies our discussion, but clearly the one-to-ten scale is more discriminating. Z probably is able to (and wants to) use a scale that has more than four categories, with the objective of hiring a "nine" or a "ten" rather than just a "good" applicant.

If you have not already done so, you should think about how you would work with the one-to-ten scale. You will find that it makes the problem much richer and offers the possibility of more subtle strategies. One might, for example, accept only a "ten" in the first two applicants, a "nine" or a "ten" in the next two, an "eight," "nine," or "ten" in the fifth applicant, and so on.

(Notice we are talking about a "ten" even though Z has apparently never met one. We would cope with this problem by suggesting that Z should change the scale so that a "ten" represented the best applicants that he had met.)

The finer resolution, however, requires more information. In particular, Z must be able to tell us the distribution of ones, twos, threes, and so on (up to tens) in his past experience. In return, it makes it possible to generate more informative output. We could, for instance, expand the horizontal axis in Figure 9.2 to show all ten categories.

If Z is unable to tell us, even roughly, the distribution of ones to tens in the population of applicants, or if Z is perfectly happy with the poor-to-good scale, then obviously there is no need for the more sophisticated approach.

Drawing on Z's Past Experience

If choosing a scale was the first breakthrough in the analysis of Z's problem, relating that scale to his past experience was the second breakthrough. We did this by summarizing Z's past experience in a distribution.

Suppose, however, that Z was brought to trial after firing the very first secretary that he had hired under false pretences. What could we do then?

The problem here is that Z has insufficient experience. He cannot provide a distribution with any confidence, but, one way or another, we would still need to decide how the categories are distributed. Perhaps the best we can do is use Z's limited experience to generate a plausible distribution. If so, it would be essential to check how *sensitive* our results were to small changes in the assumed distribution! Alternatively, we could interview secretaries at several large employment agencies and build our distribution from that experience. However, that too would involve an assumption, namely, that secretaries who respond to selective advertisements in the newspapers are no different from secretaries at employment agencies!

Different Data Lead to Different Approaches

Notice how our approach to this problem depended on the available data and the assumptions we made. For example, we noted that at any stage in the interview process, Z would know all about those applicants that had been refused, but we never made any use of this.

Suppose, however, that you had known there were going to be eighty rather than eight applicants. Would you have used this in-

formation then? How does the number of applicants influence your approach?

With eighty instead of eight applicants we would almost certainly have wanted to be more discriminating and so would definitely have used the one-to-ten rather than the poor-to-good scale. Also, we could have used information about the earlier applicants to improve our decision making with the later applicants. For example, we could have modified our distribution on the basis of what Z learned during the first 20 or 30 interviews. Of course, we would still have advised Z to hire the first "ten" he interviewed.

Similarly, if the judge had insisted on interviewing the applicants and scheduling the interviews himself, we could not have assumed that the applicants were scheduled randomly. The purpose of our model then would be to outwit the judge. For example, if the first applicant ranked "nine" on Z's scale, the second "eight" and the third "seven," we would probably want a strategy that took cognizance of that trend. The type of model would then have been completely different.

Implementing the Model

We have not discussed how to implement our model on the computer. The details would depend on how competent a programmer you are. You could, for example, just use the random number generator on a spreadsheet to generate hundreds of sets of tables such as Table 9.1, and then you could test various strategies by hand. Alternatively, you could build the strategies into computer programs and get the computer to accumulate and use the information for computing risk.

Is There a "Best" Strategy?

We like the fact that we did not actually choose a strategy for Z, but rather gave Z sufficient information (in the form of histograms such as Figure 9.2) to choose a strategy himself. The advantage of this is that we do not have to make any assumptions about Z. For example, we do not need to know whether Z is a compulsive gambler who wants to maximize the chances of hiring a "ten" irrespective of the associated chance of hiring a "one" or whether Z is cautious and conservative. Figures such as 9.2 are a good way of presenting results because they make it easy for Z to look at and consider trade-offs.

However, there is a way of defining the best strategy if we ignore Z's personality and assume that Z is coldly rational. Can you think of what that might be? (It may be the way you have approached the problem; it could even be that thinking about the analogous problem of blackjack led you to this kind of approach!)

Suppose Z is interviewing the third applicant who is a "six" on the scale of one to ten. Using our model over and over again, we could compute the probability that Z will find and say "yes" to an applicant with a score of more than six during the remaining interviews. (Think about how you would do this.) If that probability is greater than .5, i.e., if Z has a more than 50 percent chance of finding a better secretary, then we would advise Z to say "no."

Think about what kind of computing you would have to do to present Z with the full details of this best strategy and how in fact you would present it.

Can you think of a more efficient algorithm?

✓

The trick here is to work backward. On the scale of one to ten, when should Z say "yes" to the seventh applicant?

That depends on the distribution. For the sake of argument, let us suppose that the distribution is uniform, i.e., there are an equal number of ones, twos, threes, and so on, up to ten, in the general population of applicants. Then the answer to the question would be "If the seventh applicant ranks 5 or above." If the applicant is below 5, there is a greater than 50 percent chance that Z will do better in the eighth (and last) interview.

Now, what about the sixth applicant?

Suppose the sixth applicant scores x on the one-to-ten scale. If Z says "no" to the sixth applicant, then there are two ways he could end up with a worse applicant:

1. He could say "no" to the seventh applicant and end up having to say "yes" to an eighth applicant with a ranking that is less than x.

He will only say "no" to the seventh applicant if that applicant's score is less than 5. The probability of this is 5/10. The probability that the eighth applicant will score less than x is $x/10$. It follows that the probability of both events (saying "no" and then losing out) is 5/10 times $x/10 = x/20$.

2. Alternatively, he could say "yes" to the seventh applicant (if he uses the best strategy, the probability for that is 5/10) but still choose somebody worse than x. The probability of the latter happening is $(x - 5)/5$, and so the probability associated with both these events is 5/10 times $(x - 5)/5 = (x - 5)/10$.

It follows that the probability that either 1 or 2 above will happen is the sum of their individual probabilities, that is

$$\frac{x}{20} + \frac{x-5}{10} = \frac{3x-10}{20}$$

so Z should say "no" to the sixth applicant only if

$$\frac{3x-10}{20} < 0.5$$

or

$$3x - 10 < 10$$

which gives

$$x < \frac{20}{3}$$

In practice, therefore, Z should say "no" if the score of the sixth applicant is six or less.

We leave you to draw a tree (look back at Chapter 5) and develop an algorithm to work out the so-called best strategy for the fifth, fourth, third, second, and first applicants.

Validation: Can Z Decide Whether Your Advice Is Good?

How could Z decide whether you had given good advice? This is not an easy question to answer, and we suggest you think about it before you read our discussion.

✓

Normally, when one has built a model one sets out to test or *validate* it after making sure that there are no "bugs" in its construction. There are usually two direct ways of validating a model. The one is to do an experiment, which means collecting new data and checking the model against it. The other is to hold back some of the data used for constructing the model with the specific objective of using it later to test the model. There are also indirect ways of checking model output, such as looking to see whether basic laws, such as the indestructibility of matter or the conservation of momentum or conservation of electric charge have indeed been satisfied.

Validating a model is often difficult to do. Sometimes one has to just use a model on the basis that it is better than no model at all, and each time evaluate how well it performed, and thus slowly build up confidence in it.

Validating a stochastic model is even more difficult.

Why?

Z cannot test our model by experiment. First, to do so, Z would have to repeat the advertisement and would run foul of the law again. Moreover, Z would have to repeat the advertisement on a fairly large scale. It would not be sufficient to solicit eight new applicants, schedule them at random, and then test the most appealing strategy, for exactly the same reason that it would have been silly to test that strategy on only one set of simulated computer applicants. Similarly, Z cannot slowly build up confidence in the model; Z is only going to use it once!

What then can Z do? If Z had kept a detailed diary, with full particulars about previous applicants, we could devise an experiment, one in which we drew eight applicants at a time from Z's diary and tested our best strategy on them, over and over again. We could compare the results of that experiment with, for example, a histogram such as Figure 9.2 for the best strategy.

If Z has not kept a detailed diary, then all we can do is carefully check our assumptions and logic. For that, after all, is what our model is all about. If Z is comfortable with our assumptions, then the logic of our arguments should convince Z of the validity of our advice.

Notice that even after the real interviews in the judge's chambers, Z cannot comment on whether our advice was good. For that one set of interviews, the actual outcome is a matter of chance; the only purpose of our advice is to bias the outcome in Z's favor.

If this is not obvious, suppose that we offered you the following opportunity: we will toss an unbiased coin only once; if it comes up heads, we will give you $100, but if it comes up tails, you must give us $2. Would you take up that offer? And if you lost, would you blame chance or the "model" that you implicitly used when you decided to take the risk?

DISCUSSION OF THE MODELING EXERCISE

Identifying Key Concepts

In this exercise we tried to help you identify key concepts. We did not want to force you into developing these concepts in the same way as we developed them, but we did want to make sure that you recognized their importance. Without them, it is unlikely that you could have reached a viable solution:

- Without a *scale* or set of categories, you have no way of comparing applicants and cannot determine whether your client's objective has been met.
- Without a *strategy,* your client will not know how to make a decision.
- Without the concept of *risk,* you have no way of comparing strategies.
- Without a *model,* you will not be able to calculate risk.
- Without the idea of a *distribution,* you have no way of building a model.

Asking Questions and Invoking Heuristics

You might have come up with the above or similar concepts in your own way, in which case you should think about how you decided these were the key concepts; how did you get there?

We used thought-provoking questions to help us, and in the process of doing so called on a number of heuristics. Let us review some of the questions we asked and the heuristics we invoked.

One of our first questions was "What information do we have?" A good heuristic is to start by reviewing what you know. At first, in this problem, we knew very little; it was important to recognize this and to identify the need to make the most of whatever we knew. Later, after interviewing Z, we found that we knew a lot, perhaps too much, and it then became imperative to sort out what aspects of that knowledge were really important for the problem we were trying to solve. That led us to force Z to condense all sorts of information into a scale of one to ten or even poor to good. Whether we have too much or too little data, asking the question is useful.

Our next question was "What assumptions can we make?" and asking this question early on is also a good heuristic. Both the assumptions we made were crucial:

1. If the judge had previously met all eight applicants, we would have had to assume that he was impartial or else advise Z's lawyer to insist on some sort of random process for organizing the sequence of interviews.

2. If the eight applicants were in some way different from previous applicants, we would not have been able to draw on Z's previous experience.

Our next heuristic was to look for a simpler version of the same problem. This was not a fruitful heuristic in this case, but it did help to make us aware of the need to rank applicants.

Unsuccessful heuristics can sometimes still be helpful because they make you think in an unusual way about your problem!

We next asked how Z could convey information about the applicants over the phone. This is a useful "trick" whenever one is trying to establish how a person perceives something. Asking them what they would say over a phone encourages them to think about their perceptions. Asking yourself how you would use what they tell you, forces you to think about what information you really need. In this case, the question focused attention on the need for a scale; if we had missed that need in the previous question, we might have caught it here.

"Imagine the solution" is a powerful heuristic we have met before. In this case it was particularly powerful because it brought out the need for a strategy and separated stupid strategies (such as: Hire the third applicant) from potentially useful strategies (such as: Hire the first "good" applicant).

It was also important to recognize that any solution to this problem was bound to be a gamble for Z, because thinking of it as a gamble leads naturally to the idea of computing risk. If we had not recognized this, we would have had no way of evaluating the different strategies. Thinking of other gambling situations is just another version of the heuristic: Can you think of an easier problem? Thinking about a game of chance, such as blackjack, helps to focus attention on risk and how one might estimate it. With a little imagination you might even have invented a gambling situation that was much closer to Z's problem.

For example, suppose your objective is to score a high number on a die. Suppose you have three throws and can stop whenever you like. If you choose to stop after the first or second throw, your score is the number on the die at that throw; otherwise your score is the number on the third and last throw.

This problem has all the essential features of Z's problem, but is easier to think about. It focuses attention, for instance, on the concept of a distribution. It also makes it easier to think about the characteristics of a "best" solution.

Questions and Heuristics: The Metalevel

Asking questions is a crucial part of modeling and problem solving. Notice that the most useful questions are those which are pragmatic; questions that do not just ask for information but ask for information in a *useful* form or ask what you would *do* with it if you had it.

Asking questions is also a crucial part of teaching modeling. Our objective is not to tell you how to build a model, but to ask you questions that will help you work out for yourself how to build a model. This is not easy to do, because we do not have any feedback from you and because we do not want to ask questions in a way that imposes our solution on you. The first difficulty is overcome when a teacher asks questions in the classroom (provided the teacher really listens to the answers), but the second difficulty is always there; the teacher has to be careful not to preempt the answers by asking questions in a certain way.

Perhaps the most useful approach to asking questions is to try to ask the questions that the student of modeling should be learning to ask himself or herself. The questions we have asked you, the reader, in this exercise are the questions that we have asked ourselves about the problem.

In the same vein, the heuristics we have suggested or discussed are those which we would have explored ourselves. Notice that heuristics tend to be most useful in the early stages of solving a problem. Heuristics such as: Look at the data, Try to feel the problem, Anticipate what a solution would look like, Think about how you would solve a simpler version of the same problem, and so on, will always be useful, even when they do not directly apply to the problem in hand, because they encourage you to think fruitfully about the problem.

Was This a Good Problem?

As always, that depends on what we wanted you to learn from it. We had a number of objectives.

First, we wanted you to learn to recognize when a strategy was needed, how to formulate strategies, and how to build models to test and compare them. This problem revolves around questions of strategy and exposes you to situations where you naturally learn to think in strategic terms.

Second, we wanted you to learn to construct links between the real world and your model world. In previous exercises we have thought of Occam's razor as a kind of filter, keeping the irrelevant detail of the real world out of our model world. In this problem we have so little information that you are forced into forging or constructing a bridge from the real world to your model world.

How did you do that?

By making assumptions. The third objective of this exercise was to expose you to the necessity of making reasonable assumptions. This is a good problem because it cannot be solved without assumptions (such as assuming that Z could draw on his past experience).

Fourth, we wanted you to learn how to cope with both a paucity and an excess of data. This problem is instructive because there are two well-separated layers of information in it. The first layer is the one with scant information; all we know at that level are the facts about Z's past behavior and the rules for the upcoming interviews. However, there is a second layer, and that concerns Z's preferences; as we mentioned earlier, we could probably go into a great deal of detail about what Z would hope to find in a secretary. There are undoubtedly other layers as well; for example, we might want to look for information in detail about the kind of person who responds to advertisements such as Z's. Normally, these different layers of information are not so clearly stratified, and one has difficulty in separating what one really needs for the model. This was a good problem because it rewarded us if we concentrated on the first layer (fleshing it out with assumptions) and cut through the details in the second layer.

Fifth, we wanted you to learn to improvise. The idea of a scale is an improvisation. It is a device for taking subjective information (the qualities of a good secretary) and turning it into numerical information that can be used in a model.

Finally, we wanted you to gain more experience in stochastic modeling. Apart from giving you the opportunity to recognize that you needed a stochastic model and to use what you learned in the tennis problem (Chapter 5), this exercise introduced you to the need for additional concepts. For example, it creates the need for the concept of a statistical distribution.

Z's dilemma provides a fertile background against which you can make imaginative use of a stochastic model. It gives you opportunities to present results in a number of different ways. Asking the question "How will Z know whether you have given good advice?" opens up a fruitful discussion about what one can and cannot achieve with a stochastic model.

A SIMILAR PROBLEM

The important concepts in this chapter were those of risk, strategy, and a statistical distribution. The following problem is simpler than Z's problem, but draws on the same concepts.

The Case of the Troubled Airline

Dodo Airlines has just completed a study at its baggage check-in counters at a large airport. Arriving passengers were monitored every 5 minutes over a total period of 20 hours. The results showed that

- There were no passengers with baggage to check 15 times (i.e., during 15 of the 5-minute intervals).
- One passenger with baggage to check arrived during 186 of the 5-minute intervals.
- Two passengers with baggage to check arrived during the remaining 39 intervals.
- Of these 258 passengers, 24 had first-class tickets, while the rest had economy tickets.

The survey did not monitor how long passengers stood in line at the counters, but it did monitor how long it took to process each passenger. (*Process time* is defined as the time taken to tag baggage, confirm the reservation, select a seat and prepare a boarding pass.) Of the 258 passengers, 12 were processed in 5 minutes, 201 were processed in 10 minutes, and the rest were processed in 15 minutes.

The survey also asked the passengers for comments; 133 of the 258 passengers complained about the service, one passenger had a "heart attack" when asked about the service, and the remaining passengers were in such a hurry to catch their planes that they refused to comment.

You see, Dodo Airlines has only two check-in counters. There are no separate facilities for first-class passengers.

Dodo Airlines is also experiencing financial difficulties. The company views check-in facilities as an unwelcome expense, but is willing to consider suggestions. Their main competitor has been running advertisements asking, "Can Dodo get you off the ground?"

Your task is to suggest several alternatives to the company. Prepare an analysis for each alternative that will enable the company executives to make comparisons.

CHAPTER **10**

THE LIBRARIAN'S DILEMMA

AN AUTHOR'S DREAM

He was in a supermarket, presumably shopping, but he did not have a shopping cart.

The supermarket was very full, and he had this odd feeling that everybody was staring at him.

"Why doesn't somebody ask him?" he heard. And then everybody was asking:

"Didn't you write the book on problem solving?" or "Didn't you write *that* book?" or "Is that your picture on the front cover of...?"

He turned around, and sure enough, there was *his* face staring back at him, smugly, from the covers of all the magazines in the newsstand. Even though he was myopic, he could read the titles of the lead articles, titles such as "Modeling for the Millions," "A Revolution in the Classroom," "A Good Heuristic Is: Read This Book!" and "The Book Computers Are Buying Their Owners for Christmas." Somehow the supermarket had turned into a public library. The librarian was besieged by a mob, chanting "We want ______ ." There were even some computers in the crowd, or were they robots, or was the librarian a robot...?

He woke up...

...and realized he still had to write Chapter 10!

A LIBRARIAN'S NIGHTMARE

Suppose it was not just an author's dream. Suppose this book was suddenly (and unexpectedly) successful. We would be surprised, but our publishers would be even more surprised. After all, they had printed only a few thousand copies. Those would be sold out months before they could print another edition. (Publishers, or at least scientific publishers, have slow reaction times.) Since the voracious public would be unable to find the book at the booksellers, they would besiege the libraries.

The libraries would have only one or two copies on their shelves. Moreover, many of the would-be readers would discover that this is not the book they thought it would be. Imagine the librarian: twenty people on the one side shouting "When are we going to get that book?" and three dissatisfied people on the other side complaining:

"This book was a waste of time."
"I was looking for a good Whodunit."
"Why didn't you tell me I wouldn't like it?"

If only the librarian had time to talk to the borrowers, to help them decide whether this was a book they really wanted to read!

YOUR PROBLEM

Perhaps the end of the dream was not so farfetched. After all, computers can communicate with borrowers via a keyboard and a monitor. Why not think of a computer as a robot librarian? Why not program it to give borrowers individual attention and advice?

You have been hired to design the program for a pilot demonstration of how a computer can advise a borrower. This book is the example for that demonstration.

Your program should be able to question somebody (anybody) who inquires about this book, and then give appropriate advice. Remember, since the book is in short supply, the librarians want to lend it to readers who are likely to benefit from it and enjoy it.

How does a robot offer "appropriate advice"?

It could choose, in some way, from a standard list of four or five alternative recommendations. Your first task is to draw up that list.

We can imagine some snide recommendations you might want to put on the list, but let's be serious. Suppose this exercise really is a prototype for a much more ambitious advice service for the library-using public. Somebody wants to know whether a particular technical book is suitable reading. He or she consults the computer. After asking a number of questions, the computer "decides" what advice to offer. That decision is one of four or five options built into its program. What four or five options would cover most situations?

DECISIONS OR RECOMMENDATIONS

Our list of decisions or recommendations states that

1. You should definitely read this book.

2. You may enjoy the book. Why not try it when the rush is over?
3. This book is too elementary for you.
4. This book is too advanced; look for an introductory text.

What questions come to mind after reading this?
Perhaps you are wondering: "Does this list cover most situations?" or "Are the alternatives sufficiently detailed?"
Alternatively, you might be asking: "What has this got to do with modeling?"

This chapter takes us right back, in spirit, to the first two chapters. Our best description of a model (in Chapter 1) was *a purposeful representation.*

What are we planning to represent in this case?

The logical process whereby a librarian reaches a decision.

And its purpose?

To advise a would-be reader.

So we are again involved in modeling. It follows that it is important to ask questions such as "Does the list cover most alternatives" and "Are the options sufficiently detailed?" These questions help to clarify the *scope* and *resolution* of the model. They are questions we need to ask in going from the real world to our model world.

We could quite easily expand the list of decisions. We could add recommendations such as "This may be a little too advanced for you" or "Look for a much more elementary book" or even "This is not a romantic novel."

But would it be prudent to do this? Why (or why not)?

Remember Occam's razor? Remember why we wanted to keep the number of variables in our models to a minimum? Expanding our list of decisions is analogous to introducing new variables. Each decision added to the list complicates the logical model we are going to build. We should therefore try to be parsimonious.

Even if we are sure that we will need more decisions later on, it is a good idea to start with a short list of decisions. People with experience in this type of modeling (they call themselves "knowledge engineers") use a heuristic called rapid prototyping: first build a model using a short list that covers most options in a general sort of way, then modify and refine it later on.

So, if your decision list is too long or detailed, prune it or modify it. Then move on to your next task.

WHAT INFORMATION DO YOU NEED?

What information do you need to help a would-be reader reach a decision? What questions should the robot librarian ask? Your next task is to list the information you need to obtain from the user.

A good heuristic is: Imagine a telephone conversation with a stranger; what questions would you ask if the stranger expressed an interest in this book and wanted to know whether to read it?

Since the robot that is going to "ask" these questions will not be able to understand the answers, it is a good idea to attach a list of potential answers to each question. For example:

Question: What is the most sophisticated book you have ever read? Is it at the level of

Answer: 1 *Principia Mathematica?*
2 *Godel, Escher and Bach?*
3 *The Adventures of Tom Sawyer?*

Draw up the list of questions (with answers) you would like to ask.

Do not worry too much at this stage whether you have redundant or missing questions. You can always go back and review your list after you have tried to use it. But each time you think of a question to ask, first ask yourself why you need that information.

✔

QUESTIONS AND ANSWERS

What attributes would you expect a reader to have if he or she is going to make good use of this book? This is a question you should have asked yourself in order to get your own ideas organized before you started drawing up that list of questions for the robot to ask would-be borrowers.

We decided that there were four general points to consider:

1. Interest
2. Experience
3. Working habits
4. Urgency

We then considered what we were looking for under each of these headings and compiled a list of questions and answers.

Interest

The main feature of this book is that it requires work on the part of the reader. We want to make sure that a potential reader is interested enough in the subject to do that work.

It is not good enough to ask "Are you interested in problem solving?" Different people might have very different ways of interpreting what we mean by "problem solving." We need to ask a question (or questions) that are unambiguous and objective, but are indicative of a person's interest in this type of book.

We decided to ask two questions:

Question 1: You are planning a visit to a friend who has moved into a new home in a town about 100 miles from where you live. Do you
Answer: 1 Look at a map and plan the best route to drive?
2 Ask your friend for rough directions?

Question 2: A salesperson suggests you need extra insulation in your attic to save on energy bills. Would you
Answer: 1 Insulate if you could afford it?
2 Sit down with the salesperson to calculate how much energy you would expect to save?

Experience

First, we want to check that a potential reader has the right background for this book. We therefore need to ask about the reader's mathematical and computational experience.

How much experience does this book really require? We would suggest at least algebra and at least a willingness to learn to use a computer. Our questions are:

Question 3: Have you taken a course in algebra?
Answer: 1 Yes
2 No

Question 4: Which of the following best describes your computer experience?
Answer: 1 I know at least one programming language.
2 I cannot program but have used standard software (such as spreadsheets or word-processing software).
3 I have not used a computer but am anxious to do so.
4 I have not used a computer and hope I never will.

Both these questions test to see whether the potential reader has *sufficient* experience, but we also want to identify somebody who has *too much* experience, the sort of person who will find this book elementary and perhaps boring. This consideration led us to the question:

Question 5: Have you ever built and used computer models?
Answer: 1 Never
2 Used but never really built
3 Have done both, extensively

Working Habits

Here we want to look for a reader who will not take the easy way out and read ahead instead of performing the tasks we set in this book. Again, it is not useful to ask "Are you conscientious?" It is less subjective to ask a concrete question that is indicative of this quality. Our question is:

Question 6: When you work on a puzzle, such as Rubick's cube or a crossword puzzle, do you
Answer: 1 Persevere doggedly?
2 Give up easily?

Urgency

Finally, since our scenario is one where this particular book is in short supply, we want to test whether it will inconvenience the borrower to wait for the book:

Question 7: Do you expect that the skills you hope to gain from this book will
Answer: 1 Be of immediate use to you
2 Be useful sometime in the future
3 Be of no obvious use?

DRAWING CONCLUSIONS

So far you have supplied the librarian robot with information (in the form of answers that a borrower will give to a set of questions) and a set of decisions. How will it use that information to choose one decision rather than another?

Your next task is to provide a mechanism which will enable the robot to draw conclusions.

Begin by looking for a representation that will link answers to decisions.
When you have found a suitable representation, apply it to this problem.
While you are doing this, review your questions and answers. Do you need all the information you have asked for? Have you provided too many (or too few) alternative answers? Are there gaps in the information? Also ask yourself whether you are still happy with your list of decisions.

Be critical about your representation too: Is it working well? How easy is it to modify? How easy to expand?

✔

OUR REPRESENTATIONS

What thoughts did you have when we asked you to develop a representation? What heuristics did you use?

A useful heuristic, when faced with a novel task, is to look for an analogy. You want to represent the flow of logic in reaching a decision. Is there anything that you have ever seen before that represents that kind of flow?

If you have learned to program a computer, you were probably taught to draw or understand a flowchart. If you think about it, a flowchart represents the flow of computer calculations and, where a computer has to make choices, the flow of its logic.

We are going to demonstrate three different representations in this section. Our first is very like a flowchart.

Flowcharts and Decision Trees

Recall that in the section on Questions and Answers we started with a list of the issues we wanted to address. These were

Interest
Experience
Working habits
Urgency

Without going into details, it is useful to ask how these issues will influence the decisions. As a reminder, let us list the decisions again:

Decision 1: You should definitely read this book.
Decision 2: You may enjoy the book. Why not try it when the rush is over?
Decision 3: This book is too elementary for you.
Decision 4: This book is too advanced; look for an introductory text.

Figure 10.1 is a *flowchart* that represents the broad flow of our logic.

Notice how well it links the issues to the decisions. The main purpose of the robot's program is to identify borrowers who are likely to make good use of the book. The main line of the flowchart does this. It shows, unambiguously, how we identify such a reader.

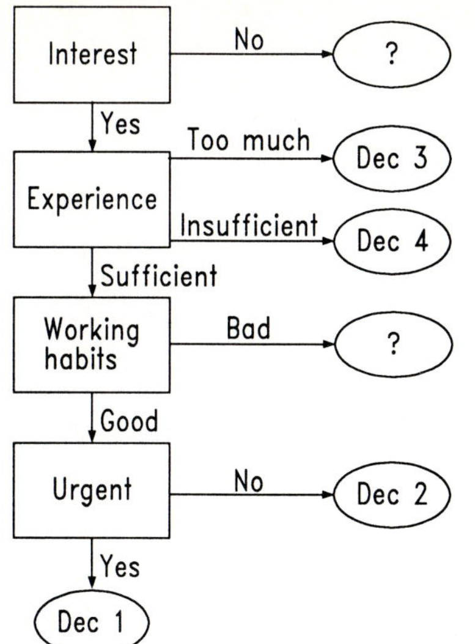

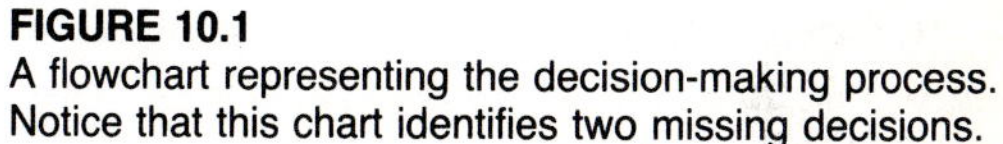
FIGURE 10.1
A flowchart representing the decision-making process. Notice that this chart identifies two missing decisions.

The side lines show how we will guide a would-be reader who does not meet all our criteria.

Notice too that this diagram identifies omissions in our list of decisions. What advice do we give somebody who is not interested in the book? Or somebody who is unlikely to work through the tasks? None of our decisions fits any of these situations.

We decided to give the same advice to both groups of people. Our new decision is:

Decision 5: You are unlikely to enjoy this book.

Figure 10.2 is the improved version of our flowchart.

Did you notice earlier on that we had left out an important decision? Perhaps you think we are really stupid not to have noticed this before. You may be right, but it is all too easy to have made omissions that look stupid in retrospect.

The important point here is that by developing our model in a general sort of way first (i.e., by using the rapid prototyping heuristic), it is relatively easy to identify gaps in decisions or issues and questions, and to go back and remedy them. Our control over the model construction is good.

Now that we have a general structure for our logic (Figure 10.2), we can go ahead and fit in our questions. This means that we have to decide how, for

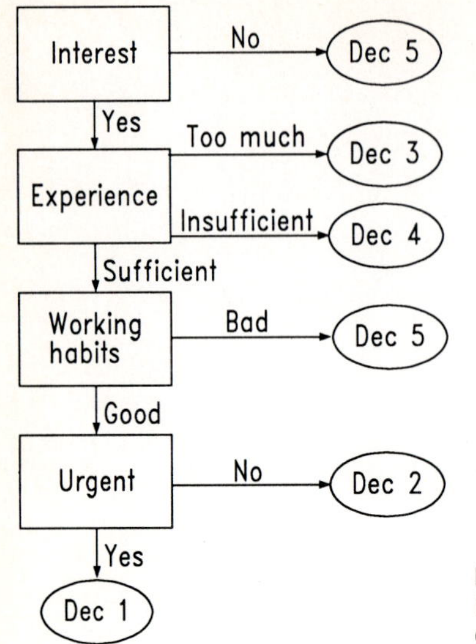

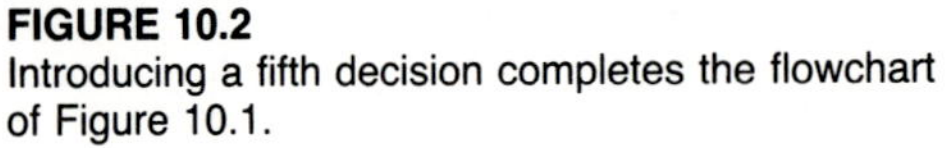
FIGURE 10.2
Introducing a fifth decision completes the flowchart of Figure 10.1.

example, we will use the answers to questions 1 and 2 to determine whether the potential reader is interested. One could argue about this, but we ended up insisting that the appropriate response to both questions was necessary to establish interest.

Fitting questions and answers into Figure 10.2 leads naturally to the tree structure of Figure 10.3. This is what we call a *decision tree*. It is a relatively easy structure to program for a computer. It is also an easy representation for humans to follow. A librarian might want to pin it next to the telephone.

Look carefully at Figure 10.3. First, does it match with the logic of the flowchart (Figure 10.2)? Next, does it interpret interest in the way we chose? Finally, what assumptions are built into it? How, for example, does it establish whether somebody has too much experience?

How does the decision tree compare with your representation? Notice that a robot (or librarian) using the tree would not necessarily have to ask the borrower all the questions. The questions asked depend on previous answers. Did your representation have this advantage?

Decision Tables

Suppose you had not noticed the analogy between the flow of logic in this model and the flow of computations in a computer program. You might then have come up with an altogether different representation.

Perhaps your thinking was: I am trying to match a set of issues against a set of decisions. A good way to make a two-way match is in a matrix or table.

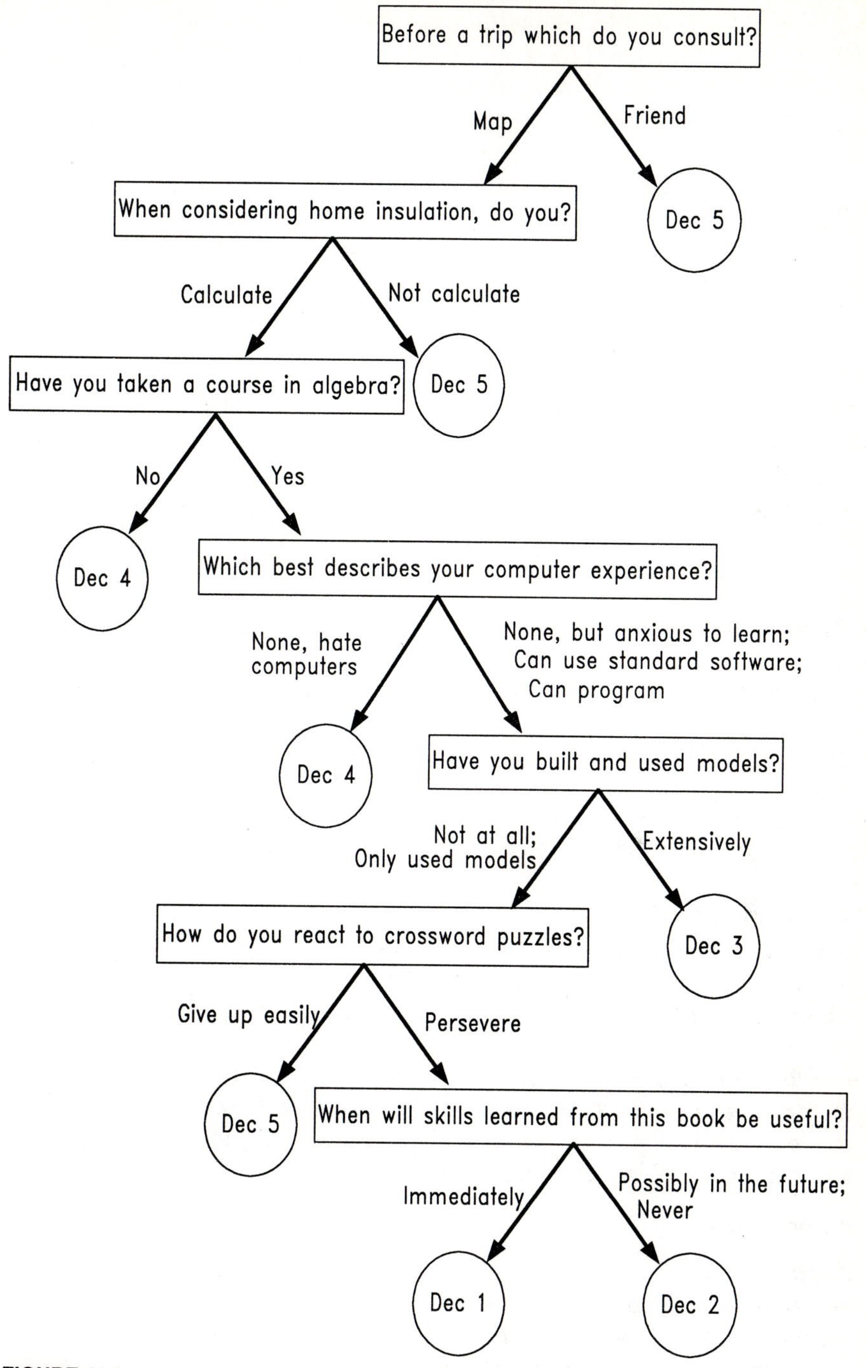

FIGURE 10.3
A decision tree constructed by adding more detail (with specific questions) to the flowchart of Figure 10.2.

Figure 10.4 is a decision table for the general structure of the librarian's model. We have a column for each of the issues, and a row for each of the decisions. Looking across a row, we mark T if a particular issue has to be true for that decision, F if it has to be false, and X if it is irrelevant.

Notice that we have to have two rows for decision 5, because it can be "triggered" by two different sets of conditions. This could be a major disadvantage in a more complicated problem where there are many different routes to a decision.

In Figure 10.5 we have replaced the issues by the questions, and instead of filling the table with F's and T's, we have inserted the numbers of the appropriate answers. We still use an X to indicate that a question is irrelevant.

Decision tables are even easier to program for a computer than decision trees. They are an excellent representation for the robot, but which do you think a human librarian would prefer? Almost certainly the decision tree.

FIGURE 10.4
A decision table. It contains the same information as the flowchart in Figure 10.2.

	Interested	Experience Sufficient	Insufficient	Good working habits	Urgent
Dec 1 Read book now	T	T	F	T	T
Dec 2 Read book later	T	T	F	T	F
Dec 3 Need more advanced book	T	F	F	X	X
Dec 4 Need more elementary book	T	F	T	X	X
Dec 5 Not your kind of book	F	X	X	X	X
	T	T	F	F	X

	Q1	Q2	Q3	Q4	Q5	Q6	Q7
Dec 1	1	2	1	1,2,3	1,2	1	1
Dec 2	1	2	1	1,2,3	1,2	1	2,3
Dec 3	1	2	1	1,2,3	3	X	X
Dec 4	1	2	1,2	4	X	X	X
	1	2	2	1,2,3,4	X	X	X
Dec 5	1	2	1	1,2,3	1,2	2	X
	1,2	1	X	X	X	X	X
	2	1,2	X	X	X	X	X

FIGURE 10.5
The decision table fleshed out with questions. This is equivalent to the decision tree in Figure 10.3.

Production Rules

Most computer languages contain a conditional construct, like the IF statement in BASIC, FORTRAN, and Pascal. If you have ever written a program using an IF statement, what purpose did it serve? What was its role in the program?

Essentially, it controls the flow of logic in the program. But that is precisely what we are trying to do here! Perhaps we can use IF statements to forge the links between issues and decisions. How could you do that?

The syntax of the IF statement is

IF [condition] THEN [action]

The condition is a logical expression (computer scientists call it a *Boolean expression*) that is either True or False. The action is an instruction to the computer. For example, one might have the statement

IF $n = 4$ THEN write ("message")

The variable n is either equal to 4 or to something else. The expression $n = 4$ is therefore either true or false. If it is true, then we are telling the computer to write "message." If it is false, the computer will ignore the write instruction.

Now consider the following statement:

IF you are the sort of person who looks at maps AND makes calculations about energy savings AND you have taken a course in algebra AND you are

NOT averse to computers AND you are NOT an experienced modeler AND you persevere with problems AND believe this book will be of immediate use THEN you should definitely read it.

We can write this symbolically as:

IF [q1a1 AND q2a2 AND q3a1 AND NOT q4a4 AND NOT q5a3 AND q6a1 AND q7a1] THEN decision 1.

This has the same syntax as an ordinary IF statement. But here the truth of the condition depends on the answers to the questions the robot librarian is going to ask a borrower. And the action is the advice that the robot will give.

We call this a *production rule*. If the condition is true, we say that it "triggers" the action. If the condition is false, we say that the rule "fails."

Which pathway in the decision tree does the above production rule represent?

We are going to look at production rules as a representation for the librarian robot. We will not go into the details of how they might actually be implemented on a computer, but it is interesting to consider, in a general sort of way, how that might be done.

You could imagine a computer program capable of reading a data file that contains

- The list of decisions
- The questions and answers
- A set of production rules

The program would be capable of recognizing key words such as DECISION, QUESTION, ANSWER, IF, THEN, AND, OR, and NOT. However, the actual text for the decisions, questions, and answers would be unintelligible to it and would be treated as just so many characters to be displayed on a screen at the appropriate times.

The program would set out first to "prove" decision 1. It would search for the first rule containing "decision 1" after the key word THEN. Next it would attempt to trigger the IF part of the rule. It would do this by displaying the first question after the word IF. The borrower would type in the number of the answer, and the computer would remember it (so that it never asks the same question twice). If the borrower's answer matches the answer in the rule, the program will ask the next question after the word IF. If the answer does not match, the program will search for another rule that leads to decision 1. If all rules leading to decision 1 fail, it will then systematically search through the rules for decision 2, and so on.

To the borrower, the computer would appear to be intelligent. The questions it asks would be in some sort of logical order.

Notice that the set of rules can contain two or more rules that trigger the same decision if it makes sense to do that. It is also possible (and very useful) to invent intermediate decisions that structure the knowledge hierarchically. Examples of an intermediate decision would be "interested" or "suitable experience" or simply "must be decision 1 or decision 2" or "dec 1 or dec 2."

Now see how quickly and easily you can write down a complete set of production rules for the robot librarian.

Our set of rules is:

Rule 1 IF [q1a1 AND q2a2 AND q3a1 AND NOT q4a4 AND NOT q5a3 AND q6a1] THEN [dec 1 or dec 2].
Rule 2 IF [(dec 1 or dec 2) AND q7a1] THEN dec 1.
Rule 3 IF [(dec 1 or dec 2) AND (q7a2 OR q7a3)] THEN dec 2.
Rule 4 IF [q1a1 and q2a2 AND q3a1 AND NOT q4a4 AND q5a3] THEN dec 3.
Rule 5 IF [q1a1 AND q2a2 AND (q3a2 or q4a4)] THEN dec 4.
Rule 6 IF [q1a2 OR q2a1] THEN dec 5.
Rule 7 IF [q3a1 AND NOT q4a4 AND NOT q5a3 AND q6a2] THEN dec 5.

Would a robot using this set of rules give exactly the same advice as a robot using the decision tree of Figure 10.3?
Which do you think is the better representation?

We would argue in favor of production rules. Our reason would be that the order in which one asks questions can easily build assumptions into the decision tree that we may not notice. Building rules is more deliberate and forces one to be conscious of all assumptions.

MAKING CHANGES

In this section we ask you to concentrate on the model we have built rather than your own model or representation, but don't forget what you did: we will ask you to see how it fits with the points we are about to raise.

We asked two questions (1 and 2) to deduce whether or not a borrower was likely to be interested in this kind of book. We then decided that only the combination of responses "q1a1 AND q2a2" would be strong enough to prove interest.

That was a very arbitrary decision. In fact the two questions are not particularly clever or probing. We could equally well argue that a suitable answer to either question would be indicative of interest.
How easy is it to alter the three different models if we make this change? Go ahead and try it.

Which representation did you find easiest to alter?

We found the production rules the easiest to update. All we had to do was replace "q1a1 AND q2a2" in rules 2, 4, and 5 with "q1a1 OR q2a2" and then we had to replace "q1a2 OR q2a1" in rule 6 with "q1a2 AND q2a1."

The decision tree was also not too difficult to change, as you can see from Figure 10.6, but if you think about it, we were lucky that we did not have to ask either question 1 or question 2 *twice* in the updated tree.

Did you struggle with this?

The decision table had to be changed quite radically. Not only did we have to alter the column entries, but we also had to add several new columns. We ended up with a much larger table than the one in Figure 10.5.

A Compromise

Suppose now that we make our logic slightly more complicated. Suppose we decide to reach a compromise between the two approaches we have so far taken to questions 1 and 2.

How would you compromise?

One could argue that people who respond appropriately to only one of these two questions do not need to read the book urgently. So we need not ask them question 7, and if they satisfy the experience and working habits criteria, we will guide them to decision 2 rather than decision 1.

Now try to implement this change. First try to draw a neat and tidy decision tree.

Is your tree neat and tidy? If it is, you have either made a mistake or else you are smarter than we are!

Did you manage to draw the tree without repeating questions?

We couldn't. But we found it much easier to change the production rules. Go ahead and see how easily you can change them.

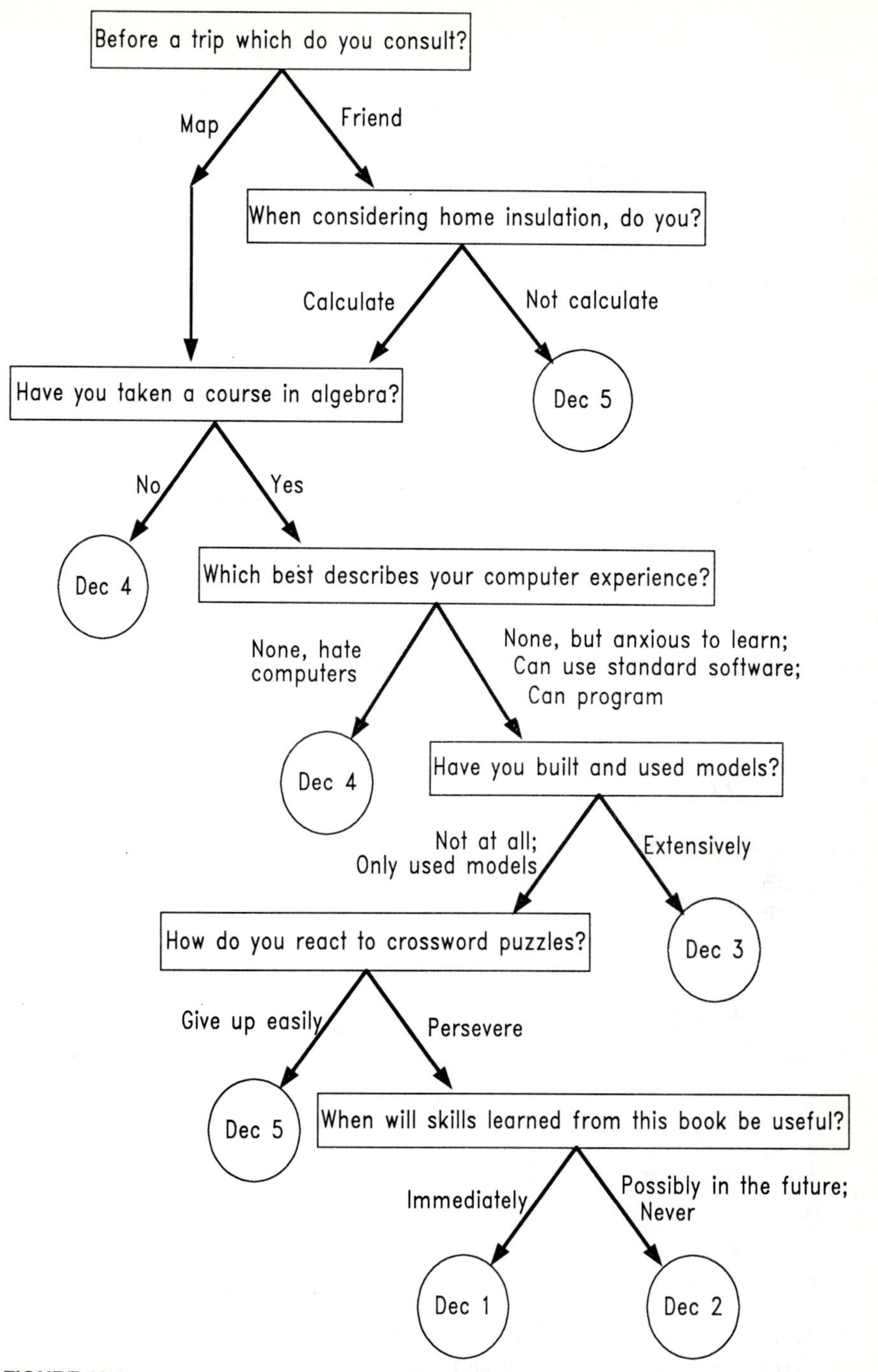

FIGURE 10.6
A modified decision tree. The logic has been changed to permit a less rigorous definition of reader interest.

We did not try to change the decision table. It could, of course, be done, but it would require patience.

Now think of how you would make similar changes in your solution to this problem.

How easily does your representation lend itself to making changes?

Finally, think of a much more complex problem. One involving, say, five times the number of questions.

Which would be your preferred representation? Why?

None of our three representations would be particularly easy to develop or modify, but we would choose production rules for two reasons:

1. We could make much more use of intermediate decisions to help structure the model.

2. The production rule representation is likely to be easier to modify than a decision table or tree.

PROVIDING EXPLANATIONS

Imagine borrowers at the library interacting with the robot we have just programmed. We can see one or two satisfied readers walking out with copies of this book. A few others would be browsing through the mystery or romance sections. One could imagine them thinking "That robot librarian really saved me a lot of wasted effort." But most of the borrowers would be lining up to sign a petition demanding the resignation of the robot.

Imagine the indignation:

"That stupid robot told me to wait. The cheek of it! Who does it think it is?"

"I agree. It asked me a question about insulation in my attic. I never said I wanted to read a book on insulation. It must have some wires crossed in its dumb attic!"

The problem is that the robot cannot explain itself. Those dissatisfied borrowers would probably have been much happier if they could have asked the robot "Why are you asking me this?" at any stage.

Suppose, for example, the robot had just asked that question about the insulation (question 2). If the borrower asked for an explanation, he or she would be far less puzzled if the robot answered back: "I am trying to determine whether this is a book that might interest you. Research has shown that this question is useful for determining whether you would be challenged by the contents of this book."

It is easy to imagine an explanation like this attached to each question in all three of our representations. If a borrower asked "Why?" the librarian or robot could respond with that explanation. But suppose the borrower wanted to know more. Where could you provide further explanations?

It is not obvious how to do this on a decision tree, but one could tag an explanation onto each rule in the production rule representation. For example, suppose the robot is trying to trigger decision 1 and has just asked question 2 in the context of rule 1. If the borrower asked "Why?" the robot could respond with the explanation attached to question 2. If the borrower asked "Why?" again, the robot could provide an explanation attached to rule 1 (such as "I am trying to establish whether you have the interest, experience, and willingness to use this book effectively"). If the borrower asked "Why?" yet again, the robot would establish that it was looking at rule 1 as a part of rule 2 so that it could provide the explanation attached to rule 2 (perhaps "This book is in short supply, and our policy is to lend it only to those who are likely to benefit from it and need it urgently").

It is harder to see how this depth of explanation could be provided with decision trees and tables. So we have yet another reason for preferring production rules: they can provide a meaningful explanation of their logic.

DISCUSSION

In this chapter we have introduced you to the idea of a knowledge model. What we have helped you to build is in fact a primitive expert system. This is a relatively new and burgeoning technology, and as such has a jargon all of its own. You have met words like "triggered" in this chapter, as well as terms like "production rules." The program we described (that would be capable of reading a knowledge model as a data file) is an example of a simple *expert system shell*. The part of the program that searches through the rules is called an *inference engine*. The data file setting out the model in an organized format is called a *knowledge base*.

A shell is rather like an empty brain: it knows how to think but has nothing to think about. Feed it a knowledge base, and it "comes alive," it has instructions for gathering information and for drawing conclusions from this information.

The example of the librarian's dilemma was chosen for several reasons. We wanted to focus on an example where we were sure our readers would share a common background. We know that our readers share at least one thing—this book. We also wanted an example that was entertaining and simple, but not trivial. It is fascinating to see how an almost trivial exercise can be used to illustrate important features of knowledge modeling in general.

We will elaborate on some of those features:

Maintaining Control

Notice how easily a simple knowledge model can burgeon and become more and more complex.

We have commented on simplicity versus complexity in virtually every chapter of this book, but the tendency to "grow out of control" is more noticeable in knowledge models than in most other types of modeling.

Why? What can we do about it?

Many of the heuristics you met in previous chapters have one important purpose: to help you prune and focus on the objectives of the modeling exercise. Use them!

In addition, there are several heuristics that we have introduced in this chapter because they are particularly powerful (and essential) in this type of exercise:

Limit the scope of the model; you can always add to it at a later stage.

Use rapid prototyping: start with a stripped-down version of the model, and work it out before you get into more detail.

Take a hierarchical approach: look at broad issues first, as in Figure 10.2, then break those issues into smaller components. (This is advice we failed to use fully in our set of production rules. We should really have first written a set of rules in terms of the broad issues such as *<u>interest</u> and <u>expertise</u>—expressed as intermediate decisions—and then written down the rules for establishing whether these issues were true or false in a particular interview.)*

As you collect information, ask yourself why you need it. This serves two purposes: it helps to weed out shallow thinking, and it paves the way for providing the knowledge model with good explanatory features.

Maintaining control of your model has important additional benefits: If you have used heuristics to structure and control your model, you will find that they also make it easier to modify the model. Moreover, other people will also find it easier to understand and modify what you have done.

Providing Good Explanations

Did you notice how difficult it is to ask for and organize the information you need in order to draw conclusions? Did you notice how difficult it is to provide good explanations?

A good explanation feature is essential. The purpose of a knowledge model is not to give advice, but rather to explain how conclusions are reached. It would be irresponsible for somebody to act on the advice of a computer program without following and understanding the underlying logic.

There is a story told about Norbert Wiener, the famous mathematician: Wiener was developing a complicated mathematical proof in class. A student asked "Isn't there another way to prove it?" Wiener thought for a while and answered "Yes." Then he thought some more and said, "And another way too."

Choosing an Appropriate Representation

We have presented three representations in this chapter: decision trees, decision tables, and production rules. There are other important representations (e.g., frames) which we have not presented but which can be found in most expert system texts, such as those referenced at the end of this chapter. Our objective was not to introduce you to all types of representations, but rather to show you how one representation might be more useful than another.

At first glance all three of the representations we introduced seemed to be appropriate for the problem. The advantages of the production rule representation only emerged as we became more sophisticated and started asking questions like "How easy would it be to make changes in the structure of the model?" and "How can we provide a good explanation feature?"

In other examples there might be structural reasons (the way the knowledge is organized) for choosing a particular representation. Or it could be that the "expert," the person whose knowledge you may be trying to represent, finds it easier to think in terms of one representation rather than another. In that case, choosing the appropriate representation could make the difference between a successful and an unsuccessful model.

APPLICATIONS

Where are knowledge models useful? What kinds of problems lend themselves to this kind of modeling?

There are a number of different answers to these questions:

Knowledge models lend themselves to problems of diagnosis, such as medical diagnosis or determining what is wrong with a car or complex piece of machinery.

They also lend themselves to situations that are more a matter of art and experience than calculation. (Think about how one might build an expert system that prompts a problem solver with heuristics that are likely to be useful!)

Knowledge models are also useful for modeling the structure and relationships between qualitative rather than quantitative variables. As an example, think of a food chain in a lake. It is very difficult (and perhaps not very useful) to build conventional models describing the interactions between all the variables (such as detritus, phytoplankton, zooplankton, small fish, big fish, fish-eating birds, and even fishing enthusiasts). It can be instructive to build a knowledge model that describes those interactions in broad terms (such as IF the pH of the water stays within a certain range, then some components of the system will thrive while others will suffer). Starfield, Farm, and Taylor describe what can be learned from building such a model (see the reference at the end of this chapter).

Knowledge models can also provide a mechanism for relating or reconciling widely different concepts and/or viewpoints. Notice how the robot librarian balanced such different concepts as "interest" and "urgency." If the robot were advising somebody to buy rather than borrow the book, it could add economics to the list of issues. It may even do some conventional calculations and tie them in with the knowledge model.

This synthesizing ability of a knowledge model can be very useful in controversial areas such as environmental planning or decision making. One of the benefits of *building* a knowledge model in this kind of situation is that it encourages a proper and constructive discussion of the issues. One of the benefits of *using* it is that the same logic is applied to all controversial issues, and that logic can be explained and defended.

Where, in any of the problems discussed in previous chapters, would a knowledge model have been useful?

✓

Remember that in Chapter 9 we sidestepped the issue of how Z would reconcile the different qualities he was looking for in a secretary, and reduce them to a simple scale such as the one-to-ten scale. We could have helped Z to build a knowledge model for doing this. A model like that could even have spared Z the embarrassment of interviewing the applicants. Somebody else could have interviewed them for him, without losing sight of his priorities!

SIMILAR PROBLEMS

We have asked students in both engineering and ecology classes to choose a problem that lends itself to this type of representation and then build a small knowledge model. The following are some of the more successful models that engineering students have built:

- Diagnosing why a car will not start
- Selecting an appropriate personal computer
- Selecting a control system for a traffic intersection
- Choosing a geophysical technique for mineral exploration
- Choosing an explosive for breaking rock

The ecology students have built models to

- Manage an aspen forest in a way that takes into account the needs of foresters as well as the black bears living in the forests

- Describe animal behavior (what should a successful grouse do next?)
- Categorize the conservation status of different rivers
- Identify different species of warblers

We suggest you make your own list of problems (that interest you) where a knowledge model would be an appropriate representation. Show the list to a friend, and explain why knowledge models fit the problems on it. Then choose one of the problems and build your own knowledge model. Pay particular attention to providing explanations in your model.

FURTHER READING

The number of new books and articles on "knowledge engineering" and "expert systems" is growing exponentially. The following is a selected sample of books and articles available at this time.

Expert Systems: Tools and Applications, by P. Harmon and R. Maus (New York, Wiley, 1988), provides an excellent overview of expert systems, advocates building small expert systems and includes a comprehensive comparison of expert systems shells and applications.

"Putting expert systems to work," by D. Leonard-Barton and J. J. Sviokla (*Harvard Business Review, 66*(2):91, 1988) discusses the improvements possible in everyday tasks through the development of small expert systems.

What Every Engineer Should Know about Artificial Intelligence, by W. A. Taylor (Cambridge, MA, MIT, 1988), provides "a practical explanation of the parts of artificial intelligence research that are ready for use by anyone with an engineering degree and that can help engineers to do their jobs better."

Expert Systems for the Technical Professional, by D. D. Wolfgram, T. J. Dear, and C. S. Galbraith (New York, Wiley, 1987), "shows you step by step how a working expert system is built, using five different expert system development tools to illustrate the wide range of expert system technology."

The paper on rule-based simulation, mentioned in the text, is "A Rule-Based Ecological Model for the Management of an Estuarine Lake," by Anthony Starfield, Brian Farm, and Rick Taylor, to be published in *Ecological Modelling* (1989).

INDEX